Materials Science and Technologies

Materials Science and Technologies

A Guide to Laser Welding
António Manuel de Bastos Pereira, PhD (Editor)
Eva Soledade Vaz Marques, MSc (Editor)
Francisco José Gomes da Silva (Editor)
2024. ISBN: 979-8-88697-613-7 (Softcover)
2023. ISBN: 979-8-89113-401-0 (eBook)

Biocomposites: Advances in Research and Applications
James R. Bush (Editor)
2023. ISBN: 979-8-89113-267-2 (Softcover)
2023. ISBN: 979-8-89113-313-6 (eBook)

Ferrite Materials and Technologies
Ravi Panwar, PhD (Editor)
Dharmendra Singh (Editor)
2023. ISBN: 979-8-89113-086-9 (Hardcover)

Perovskite Solar Cells: From Materials Science to Device Engineering
Thembinkosi Donald Malevu, PhD (Editor)
2023. ISBN: 979-8-89113-130-9 (Softcover)
2023. ISBN: 979-8-89113-188-0 (eBook)

Physics and Mechanics of New Materials and Their Applications, 2021 – 2022
Ivan A. Parinov, DrSc (Editor)
Shun-Hsyung Chang, PhD (Editor)
Arkady N. Soloviev, DrSc (Editor)
2023. ISBN: 979-8-88697-542-0 (eBook)

More information about this series can be found at
https://novapublishers.com/product-category/series/materials-science-and-technologies/

Shiv Prakash Singh
Editor

Properties and Uses of Metallic Glass

DOI: https://doi.org/10.52305/DIOG6506

Copyright Clearance Center
Phone: +1-(978) 750-8400 Fax: +1-(978) 750-4470 E-mail: info@copyright.com

NOTICE TO THE READER

Library of Congress Cataloging-in-Publication Data

ISBN: 979-8-89113-569-7 (softcover)
ISBN: 979-8-89113-631-1 (e-book)

Published by Nova Science Publishers, Inc. † New York

Contents

Preface

This book deals with the properties and applications of metallic glass. It contains the overall understanding of metallic glass, micromachining, nanostructures, functional properties of Ni-based metallic glass, and the use of computational methods. It will provide a fundamental understanding of the structure and relational properties of metallic glass.

Chapter 1 - The examination for new and advanced materials has been the major concern of the materials scientists during the past several decades. Recent surveys have focused on the development of the properties and performance of existing materials and/or synthesis and development of completely novel materials such as bulk metallic glasses (BMGs). Important enhancements have been achieved in the mechanical, chemical, and physical properties of glassy materials by the addition of alloying elements, microstructural modification and by subjecting the materials to thermal, mechanical or thermomechanical processing methods. Also, the bulk metallic glasses (BMGs) are growing high-performance engineering materials that are on the precipice of extensive commercialization. In the proposed chapter, the authors review briefly about various properties and applications of bulk metallic glass fields either theoretically or experimentally reported so far from the various literature. Such chapter creates ready reference materials for the scientific community.

Chapter 2 - The use of bulk metallic glass in recent days is finding applications in several areas ranging from sports goods to biomedical devices. Many of these applications include micro-parts made of bulk metallic glasses. Bulk Metallic Glasses are a new class of light weight metallic alloys having amorphous microstructure and superior properties than amorphous silicate glasses and many other crystalline metallic alloys. These fascinating properties make bulk metallic glasses suitable to be used in micro products. However, to fabricate complex-shaped geometries on this material as needed in various micro-parts, mechanical micro-machining is a reliable and economical process capable of creating good quality intricate micro-features.

This chapter discusses micro-machining characteristics and machinability aspects of amorphous bulk metallic glass mostly based on the experimental investigations and the developed models of the process performances. To the authors' belief, this book chapter would enable manufacturing engineers and researchers in the area of micro-machining to explore the fascinating opportunities of using bulk metallic glass in micro-devices.

Chapter 3 - Bulk Metallic Glasses (BMGs) are the newest functional glass metallic systems, having superior mechanical and thermal properties when compared to their crystalline counterparts with low critical cooling rates. The current chapter highlights the significance of bulk metallic glasses (BMG) based on [Ni-(Mo/Cr)-Si]:[Nb/Ti] systems and their distinct properties such as avoiding sulphuric acid embrittlement, overcoming strain cracking issues, preventing pitting, and enhancing corrosive resistance, all of which are obligated for aerospace and lightweight vehicle implementations. Advances in [Ni-(Mo/Cr)-Si]:[Nb/Ti] are still ongoing to improve towards industrial application with research being carried out on the aspects of corrosion properties. Thermal, mechanical, and potentiometric parameters variation in samples can be explained using microstructure changes.

Chapter 4 - Glass is a thermodynamically metastable and structurally disordered material. Metallic glass has attracted special attention for its interesting mechanical, thermal, electrical, and magnetic properties. Conventional bulk metallic glasses (CBMG) consist of two or more elements and are prepared by the conventional melt-spin technique. CBMG does not show any microstructure and is homogeneous. Recently, nanostructured metallic glass (NMG) attracted special attention for its unusual structure and other fascinating mechanical, thermal, magnetic, and electrical properties. The NMG is primarily prepared by electrochemical, chemical route synthesis, and physical vapor deposition techniques such as inert gas condensation, sputtering, pulsed laser deposition, etc. The atomic structure of the NMG is different than the CBMG. The NMG consists of two features, a glassy core and glass-glass interfaces. Due to this unique structural feature of NMG, it shows exciting properties compare to its CBMG. The current understanding of the NMG is at the preliminary stage, and it needs more studies to explore its fascinating properties. It is believed that nanostructured metallic glasses bring a next-generation revolution in nanotechnology.

Chapter 5 - Over the past several decades bulk metallic glasses have gained special recognition owing to their exceptional hardness, wear resistance, high Young's modulus, and corrosion resistance. There are several theories which explain their formation on an atomic scale. The ease of formation of metallic glassy structure has been measured in terms of glass forming ability. Further, the glass forming ability depends on various thermodynamic and topological parameters. However, it is always a difficult challenge to arrive at a bulk metallic glassy alloy with optimized chemical composition through experimental methods. Therefore, data driven computational approaches have become more powerful tools in designing optimum alloys. These approaches include artificial intelligence and machine learning methods. This chapter critically reviews the historical evolution of such methods used in development of bulk metallic glasses and their modal architecture. Finally best AI/ML models have been recommended herein.

Chapter 6 - Metallic glasses are one of the advanced structural materials having disordered atomic structures. These materials possess high strength and low ductility in comparison to their crystalline counterparts. The commonly observed deformation mechanism in metals, such as dislocation slip, twinning, or diffusion, is not observed in metallic glasses. The atomic level structure can be a combination of disordered atomic clusters plus the empty space between them, which is terms as free volume. It is the free volume region in which the atomic rearrangements occur due to the application of temperature or stress and results in plastic deformation. To date, there are two proposed deformation mechanisms in metallic glasses, i.e., the diffusive movement of atoms, and the second is the shearing of shear transformation zones (STZs). Furthermore, shear bands, which are an aggregate of STZs, nucleate and propagate to cause plastic deformation. In the above case, the deformation is observed to be heterogeneous. The following sections will present a detailed introduction, classical molecular dynamics methodology, and the deformation behavior and mechanisms in metallic glasses at the nanoscale.

Chapter 1

Bulk Metallic Glasses: A Brief Review

Aditya M. Vora*, PhD

Department of Physics, University School of Sciences, Gujarat University, Navrangpura, Ahmedabad 380009, Gujarat, India

Abstract

The examination for new and advanced materials has been the major concern of the materials scientists during the past several decades. Recent surveys have focused on the development of the properties and performance of existing materials and/or synthesis and development of completely novel materials such as bulk metallic glasses (BMGs). Important enhancements have been achieved in the mechanical, chemical, and physical properties of glassy materials by the addition of alloying elements, microstructural modification and by subjecting the materials to thermal, mechanical or thermomechanical processing methods. Also, the bulk metallic glasses (BMGs) are growing high-performance engineering materials that are on the precipice of extensive commercialization. In the proposed chapter, we review briefly about various properties and applications of bulk metallic glass fields either theoretically or experimentally reported so far from the various literature. Such chapter creates ready reference materials for the scientific community.

Keywords: bulk metallic glass (BMG), applications, properties

---

* Corresponding Author's Email: voraam@gmail.com.

In: Properties and Uses of Metallic Glass
Editor: Shiv Prakash Singh
ISBN: 979-8-89113-569-7

Introduction

In the long past of materials science and technology, which is inflated with many an attractive account of the discovery and application of new ceramic, polymeric, or composite materials, the recent advent of rapidly solidified metals, particularly the exotic non-crystalline ones called bulk metallic glasses (BMGs), as new engineering materials of fabulous promise, constitutes a unique and bright new chapters. The most noteworthy aspect of the fast progress of the rapid quenching of metallic melts, which many discriminating engineering scientists consider to be the most important revolution in materials technology of the present half of the country, lies in the fact that first experiments in this field were led out of sheer scientific interest as part of a basic research project on metals and alloys. Understanding the important aspects and the formation mechanisms of glasses remains an outstanding unexplained problem of condensed matter physics. Broadly, it is also known as disordered materials (Vora, 2004, 2008; Gandhi, 2022; Inoue, 2001, Kovalenko et al., 2001; Güntherodt et al., 1981, 1983, 1994; Elliot, 1990; Cusack, 1987; Anantharaman, 1984; Zallen, 1983).

Generally, the major things in the solid state physics have been restricted to understand the properties of crystalline solids but research on non-crystalline solids such as BMGs or amorphous solids has become one of the fashionable subjects in the field of solid state physics, for the last several years, partly because of the outstanding progress in the preparation techniques of these materials, and partly because of a promising prospects for these materials as devices. Such solids have electronic properties normally related to metals but atomic arrangement is not periodic. Bulk metallic glasses are made up of two or more than two components of metals provide us physically exciting system for theoretical research. Additionally, the BMGs play an important part in the field of materials science and engineering, which opens the door to research for both theoretical and experimental (Luborsky, 1983; Waseda, 1980; Mitra, 1976; Doremus, 1973; Rao, 2009; Basu et al., 2003).

The range of applications of BMGs is enormous and spreads from the common window glass to high capacity storage media for digital devices. Despite the abundant applications of non-crystalline materials, the basic understanding of their properties is distant from being complete. As many years before, we do not yet know very exactly and effortlessly, why certain liquids avoid crystallization under freezing below their melting temperature and stay indefinitely long in a state characterized by worldwide topological disorder. A huge amount of work has been completed in this area of non-crystalline amorphous bulk metallic glasses, both experimentally and theoretically. Because of the difficult complexity of the problems, numerical simulations using high-performance computing facilities are an essential tool in these studies. Many computational methods have been established in an effort to gain more insight into the enigmatic properties of BMGs. But, the decisive test of a model is the experiment. The attitude toward the comparison with experimental data allows us to distinguish between two abstractly different approaches connected with the computer simulation (Nu and Luong, 2016; Hebert, 2011; Hofmann, 2013; Qiao and Pelletier, 2014; Chen et al., 2019; Wang, 2012).

The first approach is to attempt the best probable quantitative arrangement with the available experimental data. Such simulations try to take into account as many truthful features as possible. This approach is definitely useful as an addition and, often, a substitute to real, costly, and sometimes not feasible experiments. Fitting the parameters involved in such models can be benefited for developing industrial technologies and improving materials. While, the second approach is the one which is frequently used in physics. A simple, often sensibly oversimplified, model is planned to capture the most prominent properties of real systems qualitatively. The use of a simple model permits us to examine the results of simulations in a most detailed way (Suryanarayana et al., 2018; Russew and Stojanova, 2013; Chatterjee, 2017).

Properties of Bulk Metallic Glasses

The BMGs in amorphous form were first produced in the laboratory by condensation from the vapor onto a cooled substrate. While Klement et al., (1960) confirmed that it is possible to produce BMG alloys by quick quenching from the liquid-quite in the same way as conventional window glass is produced by quenching from the melt. The examination of the physical and chemical properties of these BMGs proved their technological utility, which is due to the unique combination of metallic and glassy behaviour. In the last several years, many alloys have been produced in an amorphous state, and it is now widely believed that nearly all metallic liquids (except perhaps pure liquid metals) would feel a transition to a glassy state, provided that crystallization could be avoided. Whether this is probable depends at a given cooling rate on thermodynamic situations that favour the disordered (liquid or amorphous) relative the crystalline state and on kinetic situations that prevent nucleation to a crystalline structure. Generally, the glassy states are not in a thermodynamically stable state, they are only metastable state with respect to the thermodynamic ground state. BMGs are organized by diversity of procedures (Vora, 2004, 2008; Gandhi, 2022; Inoue, 2001, Kovalenko et al., 2001; Güntherodt et al., 1981, 1983, 1994; Elliot, 1990; Cusack, 1987; Anantharaman, 1984; Zallen, 1983).

1. Desertion of metallic elements in vacuum and reduction of their vapor on a chilled substrate.
2. The atoms are detached from the foundation under attack with active inert gas atoms via sputtering process.
3. In chemical or electroless deposition method, the ions in an aqueous solution are put onto substrates by chemical reactions.
4. The chemical reactions necessity for the presence of an electrical potential in electrodeposition.
5. Fast quenching from the molten state.

Generally, the BMGs are prepared by the latter method. The original usage of BMGs was for attractive purposes. Thereafter, it was used for containers and this usage is still vital today. In normal language, the word glass is commonly characteristic to an inorganic solid material that is usually transparent or translucent as well as hard, brittle and impervious to the natural elements. This usage originates from our ordinary experience with windows, decorative, wine glasses, etc. In the scientific literature, though the word glass has a much wider meaning. In fact, it is frequently used as a synonym for amorphous solid. To avoid possible uncertainty, here we mostly follow the explanations of this term available in the literature is as follows (Vora, 2004, 2008; Gandhi, 2022; Inoue, 2001, Kovalenko et al., 2001; Güntherodt et al., 1981, 1983, 1994; Elliot, 1990; Cusack, 1987; Anantharaman, 1984; Zallen, 1983):

> "A material is amorphous when it has no long-range order i.e., there is no regularity in the arrangement of its molecular constituents on a scale larger than a few times the size of these groups." (Klement et al., 1960)

Or

> "An amorphous or non-crystalline solid is the one which possesses neither the long-range translational order (periodicity) characteristic of a crystal nor the global orientation order characteristic of a quasi-crystal."

Numerous glass technicians object to the above explanation of glass. These workers make glass by chilling a liquid in such a way that it does not crystallize, and feel that this procedure is a vital characteristic of a glass. Various former authors claim this principle (Klement et al., 1960):

> "A glass is a material, fashioned by freezing from the usual liquid state, which has become more or less stiff through a substantial rise in its viscosity."

As per Jones or more concisely (Klement et al., 1960), "Glass is an inorganic creation of synthesis which has been chilled to a stiff form without crystallization." As considered from ASTM Values for Glass (Klement et al., 1960), or "Any liquid or supercooled liquid whose shear viscosity is greater than about 1013 poises is called a glass."

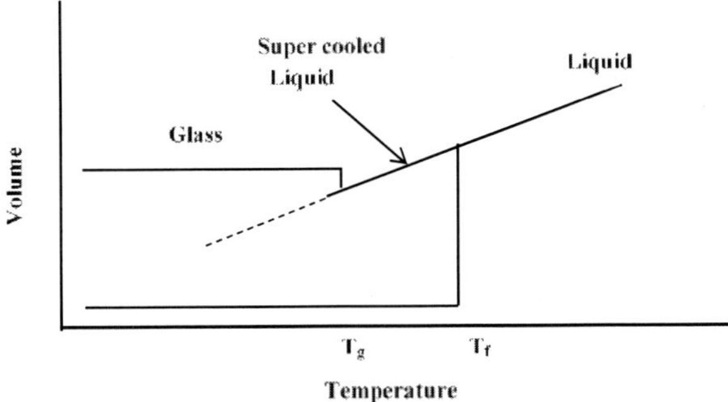

T_f = Freezing or Melting Temperature, and T_g = Glass Transformation Temperature

Figure 1. Volume-Temperature relation between glassy, liquid and crystalline states.

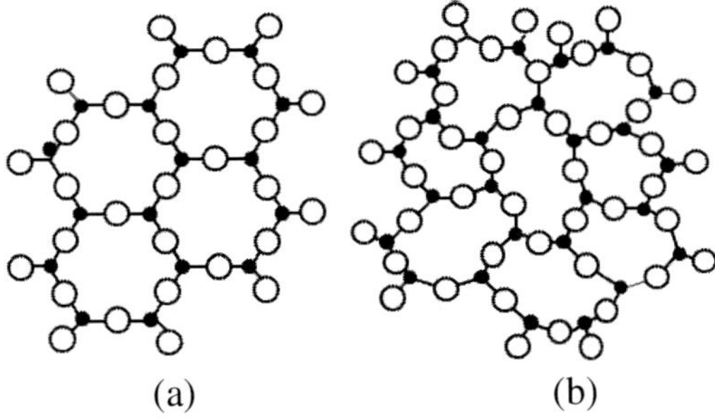

Figure 2. Schematic 2D analogs, showing the difference between: (a) the regularly repeating structure of a crystal and (b) random network of a BMGs (Vora, 2004, 2008).

This value is taken to define the liquid-glass transformation temperature Tg and to define the boundary between the liquid state and glassy state, which is shown in Figure 1 (Klement et al., 1960, Vora, 2004, 2008).

BMGs, like ordinary liquids, possess atomic order only over a range of the order of one interatomic separation are shown in Figure 2 (Klement et al., 1960).

Their high strength, bend ductility and toughness in accumulation to their soft magnetic properties, high electrical resistivity and excellent corrosion resistance make BMGs attractive materials for industrial applications (Vora, 2004, 2008; Gandhi, 2022; Inoue, 2001, Kovalenko et al., 2001; Güntherodt et al., 1981, 1983, 1994; Elliot, 1990; Cusack, 1987; Anantharaman, 1984; Zallen, 1983; Qiao and Pelletier, 2014). Generally, the BMGs have the subsequent significant features (Chatterjee, 2017):

- They are multi-component structures, which are having more than three sub-components in non-crystalline phase.
- They are formed at low solidification rates ~ 103 K/s or fewer.
- There is a huge alteration among the glass transition temperature and the crystalline temperature, herewith ensuing in a huge supercooled liquid region.

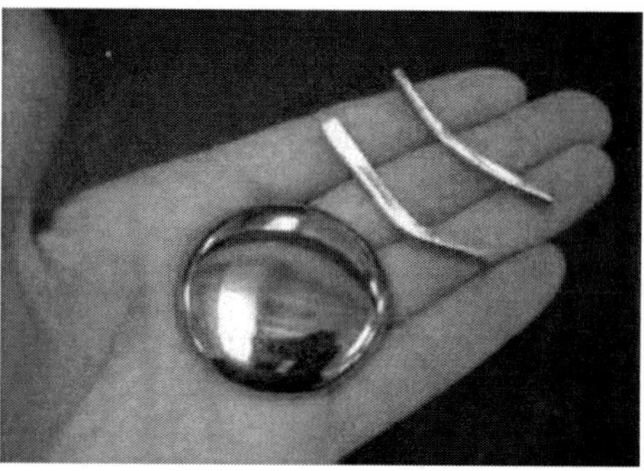

Figure 3. What do metallic glasses look like? (Rao, 2009).

In the existing era, the BMGs are the lime lighted attention of penetrating research among the metal or alloy community. Currently, the research on the non-crystalline solids such as bulk metallic glasses or amorphous solids has become one of the newest subjects in the field of solid-state physics or condensed matter physics, meanwhile the last several years, partly because of the extraordinary development in the preparation and manufacturing techniques of such materials, and partly because of auspicious scenarios for such materials as devices applications. The electronic properties of amorphous solids are apparently and usually related to metals but the atomic arrangement is not a periodic one. Here, Figure 3 shows how a metallic glass looks normally (Gandhi, 2022).

Synthesis of Bulk Metallic Glasses

In this section, some well-known processes are discussed which are used in the production of BMGs.

Vapor State Processes

Physical Vapor Deposition (PVD)

In this method as displayed in Figure 4, the BMGs to be dropped is vaporized by physical heating or by sputtering and then deposited onto a surface (Chatterjee, 2017, https://en.wikipedia.org/wiki/Physical_vapor_deposition).

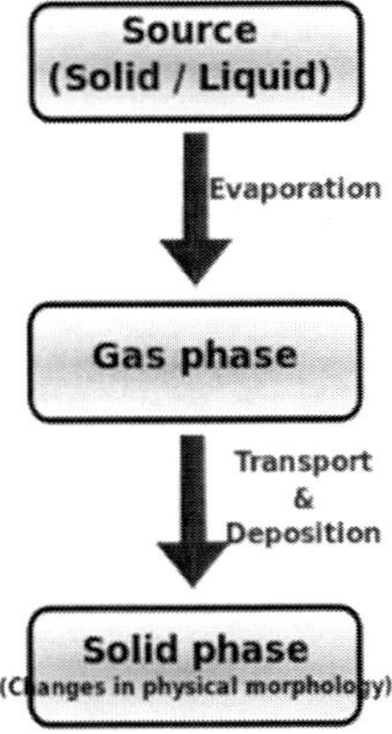

Source: https://en.wikipedia.org/wiki/Physical_vapor_deposition.

Figure 4. Process diagram of PVD.

Some of the examples of this process are:

1. *Pulsed Laser Deposition*: Here, high powered laser is used to vaporize the material.

2. *Sputter Deposition*: Here the material is bombarded with 'glow' plasma which takes away the material as vapor and then deposited on the target.
3. *Electron Beam PVD*: Material to be deposited is vaporized using a high energy electron beam in vacuum environment and then transported via diffusion to be deposited on the target.
4. *Cathode Arc Deposition*: A high powered electric arc is discharged at the target material and deposited.
5. *Evaporative Deposition:* The material to be deposited is heated to a high vapor pressure by electrical resistance heating in a vacuum chamber.

This process are having certain advantages such as (Chatterjee, 2017, https://en.wikipedia.org/wiki/Physical_vapor_deposition): (1) PVD coverings are occasionally tougher and more erosion resistant than coverings applied by the electroplating process, (2) Most coverings have high temperature and virtuous excellent abrasion resistance, impact strength and are so tough that defensive topcoats are nearly never essential and (3) Capability to use almost any type of inorganic and some organic covering materials on an similarly varied group of substrates, (4) Surfaces using an extensive diversity of textures, (5) More ecologically friendly than old covering procedures such as electroplating and painting and (6) More than one method can be used to credit a given film. Though some drawbacks are as : (1) Precise technologies can enforce restrictions e.g., line-of-sight transmission is distinctive of most PVD covering methods, but these procedures are that permit full attention of compound geometries, (2) Some PVD technologies classically work at very high temperatures and vacuums, necessitating distinct care by functioning personnel and (3) Needs a chilling water organization to disperse huge heat loads (Chatterjee, 2017; https://en.wikipedia.org/wiki/Physical_vapor_deposition).

Chemical Vapor Deposition (CVD)

Here, the aim substantial is unprotected to some chemical material which decays on/responds through it and produces the required coverage on it. It can be classified giving to the working pressure viz. ultra-high vacuum pressure or atmospheric pressure or low pressure (Chatterjee, 2017, https://en.wikipedia.org/wiki/Chemical_vapor_deposition). Some of the examples are:

1. Aerosol Aided CVD: The predecessors are transported to the target via an aerosol gas which can be produced ultrasonically.
2. Fast Thermal CVD: A heating lamp is utilized to warm the target to permit informal deposition.
3. Combustion CVD: A flame is utilized in the open air to deposit thin film coverings.
4. Plasma Enhanced CVD: Plasma is used to increase the affinity of the chemical precursor for the target. It allows for low temperature deposition and is useful especially for the semiconductor industry.

Ion Implantation

It is a lower temperature procedure through which ions of one element are enhanced into a hard target substance, thus altering the chemical, physical or electrical properties of the target material. It is utilized in the fabrication of the semiconductor device, in the metallic finishing and in the research of the materials science. The ions can vary the fundamental configuration of the target (if the ions deviate in configuration from the target substantially) if they halt and remain in the target. It also reasons physical and chemical changes when the ions enforce on the target at high energy. The crystal assembly of the target can be damaged or even destroyed by the active influence cascades, and ions of adequately high energy (10s of MeV) can cause nuclear transmutation (Chatterjee, 2017, https://en.wikipedia.org/wiki/Ion_implantation). The Setup diagram of such process is seen in Figure 5.

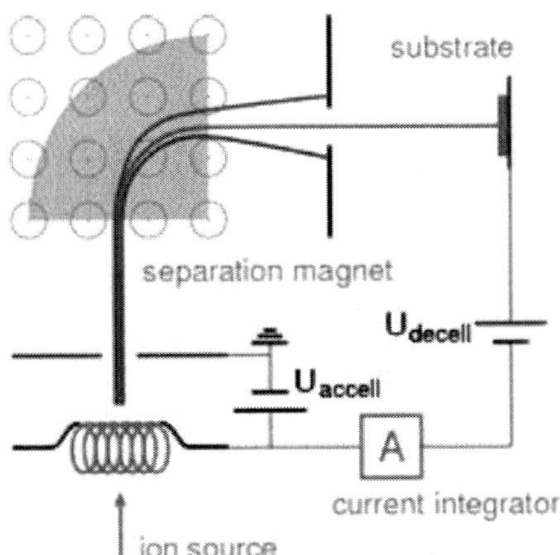

Source: https://en.wikipedia.org/wiki/Chemical_vapor_deposition.

Figure 5. Setup diagram of ion-Implantation process.

Liquid State Processes

Rapid Solidification Processing (RSP)

Here it contains quick solidification of the liquid 'glassy' melt at chilling rates ~ 106K/s to accomplish a BMG assembly. It wants that the warmth to be detached from the melt at a very high rate owing to which the unit thickness of the final product is restricted to micron ranges (Suryanarayana and Inoue, 2018; Chatterjee, 2017). The traditional approaches of attaining such high cooling rates are:

Droplet Technique

A melted metal is atomized into minor dews and then these dews are chilled either by being visible to a stream of cold air or an inert gas. Another technique of solidification is through striking such dews i.e., splatting on a decent heat leading surface (Chatterjee, 2017).

Jet Technique

In this technique, a smooth stream of melted metal is successively cold by moving it in connection with a moving freezing surface. The BMGs are molded in the form of sheets or ribbons or wires (Chatterjee, 2017).

Surface Melting Techniques

Such method contains fast melting at the surface of a bulk melt and then solidification via quick heat abstraction into the unmelted zone via laser action.

It is the most common procedure of creating metallic glasses. They are having some applications in scientifically numerous areas such as medical implants and dental amalgams, fuel cell technology, superalloys and powder metallurgy tool steels etc. (Suryanarayana and Inoue, 2018; Chatterjee, 2017). The products molded usually have very thin cross sections and therefore it is tough to discovery straight applications. Such BMGs are utilized to produce BMGs of higher cross zones which can then be utilized for a assembly of applications.

Splat Quenching

It classically includes quick solidification of liquid metal among two rollers which are cooled unceasingly to remove the heat. A thin sheet of BMGs is shaped by low volume to area ratio. Fundamentally, it is a liquid rolling process. A good method is Duwez and Willen's gun technique (https://en.wikipedia.org/wiki/Splat_quenching). In which, the liquid metal is thrown to a quencher plate due to which its portion quickly increases and it speedily chills to form a thin sheet. A wider area of close BMGs can be formed. The two significant procedure parameters for splat quenching are the velocity of the dews and their volume. If the volume is too vast or the velocity is too minor, the drop doesn't completely solidify. Therefore, it is practically determined as to what is the best drop size and velocity that is suitable for creating a thin sheet having even thickness, composition and good mechanical properties. Hence, products which are formed utilizing splat quenching typically have close glassy assembly and outstanding Paramagnetism owing to which such process is appropriate for applications associated with magnetic shielding etc. (Chatterjee, 2017; https://en.wikipedia.org/wiki/Splat_quenching).

Solid State Processes

Mechanical Alloying

It is a powder metallurgy type technique, which was established by John Benjamin in the middle of 1960's at INCO (Suryanarayana and Inoue, 2018; Chatterjee, 2017; https://en.wikipedia.org/wiki/Mechanical_alloying). The dissimilar stages elaborated in it are:

1. The crushing medium of Tungsten carbide or stainless steel balls and the mixed elemental powder particles are put in a vessel.
2. The vessel is agitated at higher speed for a fixed time period.
3. The soft powders of each metallic become crushed and adopt a flat structure with tinny cross segments. Such flat forms have a coated system.
4. Owing to weighty plastic distortion, crystal faults e.g., displacements, grain limits, positions etc. are presented in the BMGs. Also, the temperature increases.
5. Owing to the increase in temperature, the diffusion is facilitated subsequent in fraternization of the metal powders to construct the alloys.
6. Such alloys can be constructed to wanted form utilizing methods such as vacuum hot pressing, hot extrusion, hot isostatic processing etc.

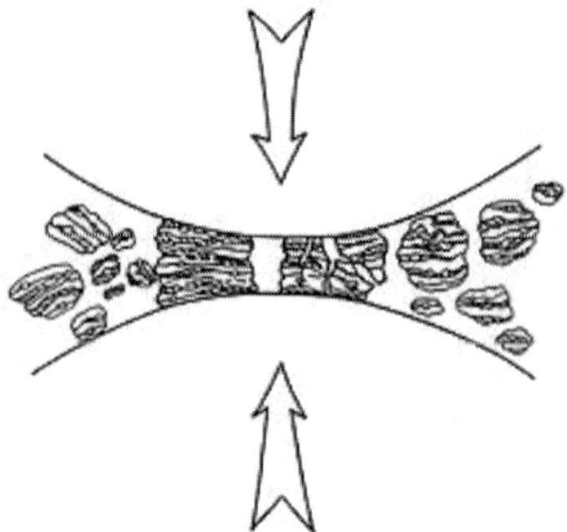

Source: https://en.wikipedia.org/wiki/Mechanical_alloying.

Figure 6. Mechanical Alloying Method.

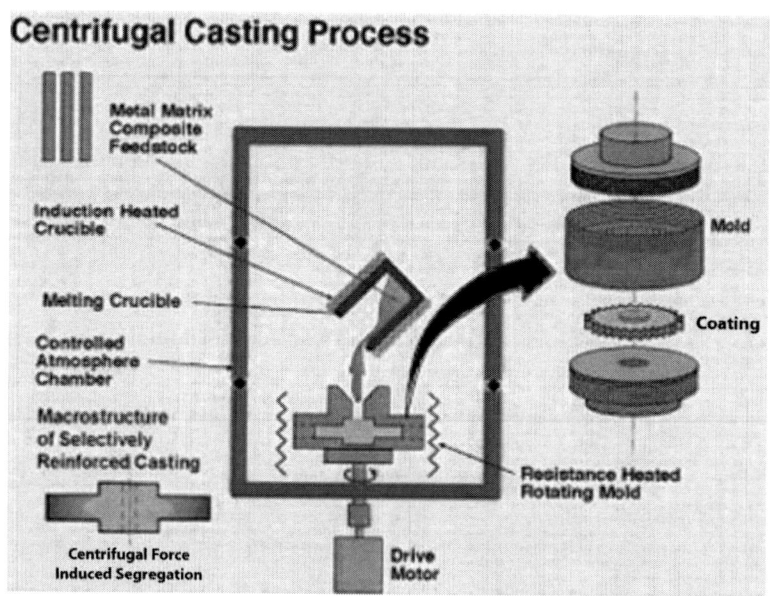

Source: http://www.themetalcasting.com/centrifugal-casting.html.

Figure 7. Centrifugal casting method.

It can also be utilized to make a diversity of non-equilibrium stages such as metastable phases, supersaturated solid solutions (SSSS) etc. (Suryanarayana and Inoue, 2018; Chatterjee, 2017; https://en.wikipedia.org/wiki/Mechanical_alloying).

Fabrication of BMGs is also carried out by centrifugal casting method (Nowosielski and Babilas, 2007). The considered centrifugal casting scheme contains two key portions: injection scheme of liquid alloy and casting device. The casting device comprises a cylindrical copper mildew, which is alternated by a motor through a belt transmission. The transmission permits to varying the rapidity of revolving mildew. Furthermore, the injection scheme contains of quartz vessel, where the alloy ingot is molten by inducing melting. The procedure of ingot melting is understood by utilizing a power supply with a higher frequency. The processes of this method is observed in Figure 6.

Figure 7 shows a schematic diagram of the centrifugal casting method with an injection scheme of molten alloy used for the production of BMG rings. Said method permits to manufacture BMG materials in the arrangement of a ring with diameter of 25 mm. The ingot in an argon atmosphere which is placed in quartz nozzle heated by induction melting and then after the molten alloy is introduced into a cylindrical copper mildew, which is revolving with rapidity of 2500-3500 rpm. The width of the ring is controlled by changing the mass of the molten alloy (Nowosielski and Babilas, 2007).

Applications of Bulk Metallic Glasses

As application arenas of BMGs is written subsequently as, (1) edge in tennis rackets, (2) striking face plate in golf clubs, (3) casing in cellular phones, (4) various shapes of optical mirrors, (5) shot penning balls, (6) electro-magnetic shielding plates, (7) casing in electro-magnetic instruments, (8) connecting part for optical fibers, (9) soft magnetic high frequency power coils, (10) soft magnetic choke coils, (11) vessels for lead-free soldering, (12) high corrosion resistant coating plates and (13) high torque geared motor parts etc. (See in the Figures 8-10).

Additionally, as object illustrations which lie at the last phase for applications, one can observe as, (1) Colliori kind fluid movement meter, (2) higher sensitivity, higher load and lesser size kind pressure sensors in numerous vehicles containing automobile, (3) slat truck cover for air plane, (4) spring, (5) in-printing plate, (6) yoke material for line actuator, (7) high density of information-storage material, (8) high-frequency type antenna material, (9) magnetic iron core for higher revolution speed motor, (10) biomedical tools such as endoscope parts etc. (Vora, 2004, 2008,; Gandhi, 2022; Inoue, 2001, Kovalenko et al., 2001; Güntherodt et al., 1981, 1983,

1994; Elliot, 1990; Cusack, 1987; Anantharaman, 1984; Zallen, 1983; Basu et al., 2003).

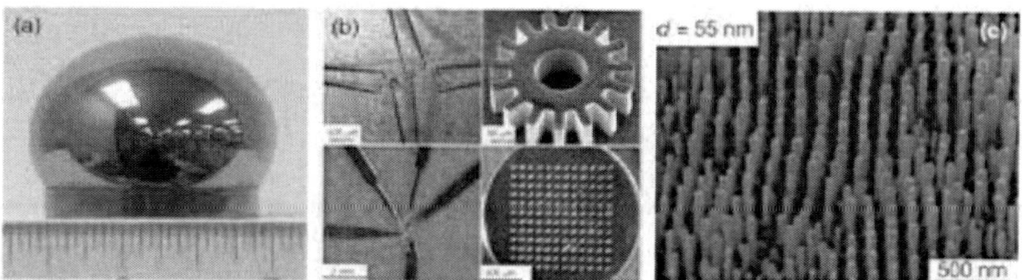

Source: http://www.materials.unsw.edu.au.

Figure 8. Biomedical and Aerospace Applications of BMGs.

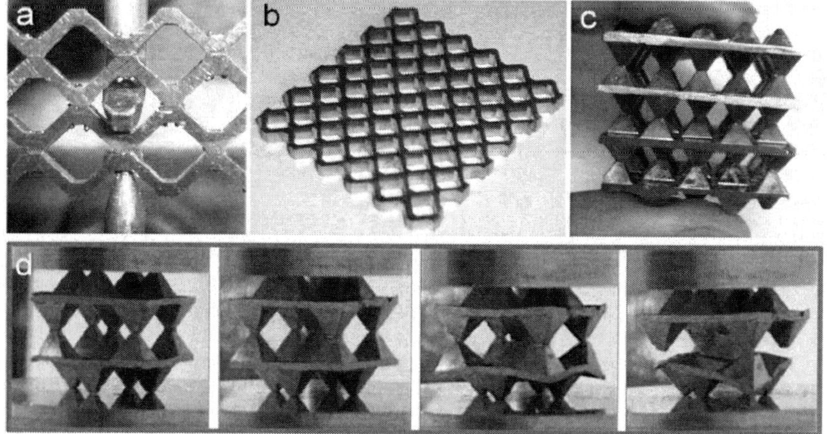

Source: http://www.wikipidia.com.

Figure 9. Some industrial applications of BMGs.

Source: http://www.eurekalert.org.

Figure 10. BMG formation.

According to their exceptional properties, following military apparatuses are being established using BMGs by the US Department of Defense (DoD) (Rao, 2009) like as sub-munition components and Fuzzes, lightweight casings for ordnance, Compound armour, thin walled casings and apparatuses for electronics, MEMS casings and components, Missile machineries: fins, nose cones, gimbals, and bodies, lighter weight fragmentation devices, aircraft fasteners, casings for night sights and optical devices, EMP and EMI Shielding. While, in future, W-reinforced BMG-composite armour-piercing warheads can substitute depleted uranium penetrators due to their similar density and self-sharpening behavior. Different most crystalline metal projectiles, which smooth on influence, the edges of BMG-composite penetrators sheer away under dynamic loading. Also, the BMGs are accurate materials for Micro-Electro-Mechanical systems (MEMS) production where low-loss coefficient, high yield strain and elasticity (and indirect good fatigue resistance) are mandatory, while low ductility and low process-zone size become unimportant at minor scale. Meaningfully for MEMS construction, the BMGs are placed as thin films and the glass forming ability (GFA) in deposition opens up a much wider range of outlines than is conceivable for BMGs (Rao, 2009).

The utmost general day to day applications of the BMGs truly kind usage of the original magnetic properties of some BMGs based on iron, nickel, and cobalt due to the absence of their crystalline assembly. In specific, the BMGs have a propensity to have low coercivity because there are no limits amid crystalline grains to block the motion of magnetic field walls and there is no magneto-crystalline anisotropy. Moreover, the resistance of BMGs to current movement is usually higher than that of crystalline compounds. Such aids to minimize eddy-current losses that happen owing to quick magnetization and demagnetization of the said material (Rao, 2009).

At the end, fine jewelry manufacturing is the newest one to support BMGs due to their unique properties, which can produce a metallic surface i.e., both enormously solid and scratch-resistant, but can be refined to a higher luster that is put over time. The BMGs also have some ugly properties such as metallic glasses, particularly large apparatuses, tend to break in fragile ways. Such nonexistence of ductility was attributed to shear localization in the BMGs. They have poor resistance to spread of fatigue cracks. They have a very higher cost compared to many general compounds. Efforts are also in progress to enlarge ductility of the BMGs. One of the methods was to encourage partial crystallization by thermal actions. This outcome in a complex structure comprising of nanocrystals isolated in the amorphous matrix. Further method was to contain second phase particles in a metallic glassy matrix creating an accurate composite structure. The particles both endorse shear band formation and inhibit shear band circulation. As shear will not be localized (which happens in the propagation of a single shear band resulting in brittle failure), ductility recovers. As current attention is raised to grow the BMGs utilizing inexpensive raw materials. It is only a matter of time previously the BMGs enter in our daily life from the current top section with higher technological applications (Rao, 2009).

Conclusion

Lastly, we summarized that the BMGs are used in numerous technological and scientific applications in the industry as well as in our daily life. It is also advantageous to build a superconductor, nanotubes, magnetic tapes, switching devices, etc. This inspires us to do research in the field of BMGs. However, the BMGs have excellent mechanical characteristics as associated with the metals. But, the main restriction is the brittle structure and lower formability owing to which it is tough to construct composite formed parts. Also, amongst the vapor phase methods, the physical vapor deposition is normally the most inexpensive and atmosphere friendly method to produce the BMGs. Each procedure has its own advantages depending on the product requirements. While it forming with vapor phase techniques, have the highest cooling rate, followed by liquid phase and then solid phase processes. Hence, the vapor phase procedures are the most expensive, followed by the liquid and then the solid phase. Therefore, in the present article, some important aspects of BMGs are narrated and discussed. By reviewing such evidence, the readers particularly the educators and the students can touch the scientific perception of the BMGs and their technological applications. There is also scope for further details of such a concept.

Acknowledgments

Computer facility developed under DST-FIST program from the Department of Science and Technology, Government of India, New Delhi, India, and financial assistance under DRS-SAP-II from the University Grants Commission, New Delhi, India is highly acknowledged by the author.

References

Anantharaman TR. *Metallic Glasses: Productions, Properties and Applications.* Switzerland: Trans Tech; 1984.

Basu J, Ranganathan S. *Bulk metallic glasses: a new class of engineering materials.* Sadhana (2003) 28:783-798.

Centrifugal Casting Process, http://www.themetalcasting.com/centrifugal-casting.html.

Chatterjee R. Manufacturing of Metallic glasses. *Advanced Materials Manufacturing & Characterization* (2017): 7:24-29.

Chemical Vapor Deposition Process, https://en.wikipedia.org/wiki/Chemical_vapor_deposition.

Chen M. *A brief overview of bulk metallic glasses.* NPG Asia Mater (2011) 3:82–90.

Cusack NE. *The Physics of Structurally Disordered Solids.* England: Institute of Physics, Adam Hilger; 1987.

Doremus RH. *Glass Science.* New York: John Wiley & Sons; 1973.

Elliot SR. *Physics of Amorphous Materials.* London: Longman Scientific and Technical; 1990.

Gandhi AL. Study of Vibrational Dynamics, Electronic Transport and Superconducting Properties of some Bulk Metallic Glasses using Pseudopotential Theory. PhD Thesis. C. U. Shah University. Wadhwancity. Gujarat. India; 2022.

Güntherodt HJ, Beck H. *Glassy Metals-I: Ionic Structure, Electronic Transport and Crystallization*. Berlin: Springer-Verlag; 1981.

Güntherodt HJ, Beck H. *Glassy Metals-II: Atomic Structure and Dynamics, Electronic Structure, Magnetic Properties.* Berlin: Springer-Verlag; 1983.

Güntherodt HJ, Beck H. *Glassy Metals-III: Amorphization Techniques, Catalysis, Electronic and Ionic Structure*. Berlin: Springer-Verlag; 1994.

Hebert RJ. *Nanocrystals in Metallic Glasses*. Nanocrystal. London: IntechOpen; 111-135; 2011.

Hofmann DC. Bulk metallic glasses and their composites: a brief history of diverging fields. *Journal of Materials* (2013). 2013:517904.

Inoue A, Hashimoto K. *Amorphous and Nanocrystalline Materials: Preparations, Properties and Applications*. Berlin: Springer-Verlag; 2001.

Ion Implantation Process, https://en.wikipedia.org/wiki/Ion_implantation.

Klement W, Willens RH, Duwez P. Non-crystalline structure in solidified gold–silicon alloys. *Nature* (1960) 187:869–870.

Kovalenko NP, Krasny YP. Krey U. *Physics of Amorphous Metals*. Berlin: Wiley-VCH; 2001.

Luborsky FE. *Amorphous Metallic Alloys*. London: Butterworths; 1983.

Mechanical Alloying Process, https://en.wikipedia.org/wiki/Mechanical_alloying.

Mitra SS. Physics of Structurally disordered solids. New York: Plenum Press; 1976.

Nu NTN, Luong TV. Potential applications of metallic glasses. *International Journal of Science, Environment and Technology* (2016) 5:2209-2216.

Nowosielski R, Babilas R. Fabrication of bulk metallic glasses by centrifugal casting method. *Journal of Achievements in Materials and Manufacturing Engineering* (2007) 20:487-490.

Physical Vapor Deposition Process, https://en.wikipedia.org/wiki/Physical_vapor_deposition.

Qiao JC, Pelletier JM. Dynamic mechanical relaxation in bulk metallic glasses: a review. *Journal of Materials Science & Technology* (2014) 30: 523-545.

Rao RB. *Bulk Metallic Glasses: Materials of Future*. DRDO Science Spectrum. 2009; 212-218.

Rowson H. *Glasses and their Applications*. London: The Institute of Metals; 1991.

Russew K, Stojanova L. *Glassy Metals*. Heidelberg: Springer; 2016.

Shunhua C, Jingyuan W, Lei X, Yucheng W. Deformation behavior of bulk metallic glasses and high entropy alloys under complex stress fields: a review. *Entropy* (2019) 21(54):01-20.

Splat quenching Process, https://en.wikipedia.org/wiki/Splat_quenching.

Suryanarayana C, Inoue A. *Bulk Metallic Glasses*. 2nd Ed. London: CRC Press; 2018.

Vora AM. Metallic Glasses: A brief review. *Materials Science Research Journal* (2008) 2:19-24.

Vora AM. Study of Vibrational Dynamics of Binary Metallic Glasses using Pseudopotential Theory. PhD Thesis. Sardar Patel University. Vallabh Vidyanagar. Gujarat. India; 2004.

Vora AM. *Vibrational Dynamics of Binary Metallic Glasses*, Germany: LAP LAMBERT Academic Publishing; 2012.

Wang WH. The elastic properties, elastic models and elastic perspectives of metallic glasses. *Progress in Materials Science* (2012) 57:487–656.

Waseda Y. *The Structure of Non-Crystalline Materials*. New York: McGraw-Hill Int. Book Com.; 1980.

Zallen R. *The Physics of Amorphous Solids*. New York: John Wiley & Sons; 1983.

Chapter 2

Micromachining of Bulk Metallic Glass: A Comprehensive Study into the Micromachining Performances

Debajyoti Ray[1,*]
and Asit Baran Puri[2]
[1]Department of Mechanical Engineering,
Sanaka Educational Trust's Group of Institutions, Durgapur, India
[2]Department of Mechanical Engineering, National Institute of Technology Durgapur, Durgapur, India

Abstract

The use of bulk metallic glass in recent days is finding applications in several areas ranging from sports goods to biomedical devices. Many of these applications include micro-parts made of bulk metallic glasses. Bulk Metallic Glasses are a new class of light weight metallic alloys having amorphous microstructure and superior properties than amorphous silicate glasses and many other crystalline metallic alloys. These fascinating properties make bulk metallic glasses suitable to be used in micro products. However, to fabricate complex-shaped geometries on this material as needed in various micro-parts, mechanical micro-machining is a reliable and economical process capable of creating good quality intricate micro-features. This chapter discusses micro-machining characteristics and machinability aspects of amorphous bulk metallic glass mostly based on the experimental investigations and the developed models of the process performances. To the authors' belief, this book chapter would enable manufacturing engineers and researchers

* Corresponding Author's Email: debajyoti_ray2002@yahoo.co.in.

In: Properties and Uses of Metallic Glass
Editor: Shiv Prakash Singh
ISBN: 979-8-89113-569-7

in the area of micro-machining to explore the fascinating opportunities of using bulk metallic glass in micro-devices.

Keywords: bulk metallic glass, micro cutting, micro milling, micro cutting forces, micro burrs

Introduction

Bulk metallic glass (BMG) alloys are light metallic alloys having amorphous microstructure and superior properties than amorphous silicate glasses and many crystalline metallic alloys. These alloys have emerged over the last three decades with fascinating promise to be used as structural or functional materials possessing attractive properties (Basu and Ranganathan, 2003). Miniature devices can be fabricated from these materials and can be used in several application areas ranging from sports goods to biomedical devices. Though aluminium alloys, titanium alloys, steel, etc. are the key materials for different micro-application areas, however, these materials have some limitations as well. Aluminium and its alloys can be die cast into near net shapes but are soft and are not enough scratch resistant. Steel is scratch resistant but denser than aluminium and therefore, has limitations in forming into small complex shapes. Titanium alloys have low density and high strength but has limitation in cast ability and poses a problem in shaping by machining. To address these limitations, and to have all the advantages available in a material, there was a need to develop a unique material. Thus in the early 1960s, the synthesis and vitrification of gold-silicon metallic alloy from the molten state using rapid cooling (of the order 10^6 Ks^{-1}) techniques by a group of researchers at Caltech, USA and further development of numerous metallic glasses in bulk form had provided a solution (Design guide 5.0, Liquidmetal, 2021). By properly designing alloy compositions, metallic glasses having a thickness much greater than 1 mm and known as bulk metallic glasses (BMGs) were developed. These metallic glasses in bulk form could be made at moderate cooling rates. Due to amorphous microstructure without having any long range of atomic order, and absence of grain boundaries and dislocation defects, this class of material possesses unique properties, such as high elastic strain, high hardness, high strength, light weight, high strength to weight ratio, high corrosion resistance in comparison to many materials. The ease of moulding into diverse shapes makes it suitable to be used for three-dimensional micro-structures. BMGs are scratch resistant and can give a high surface finish and lustre so that it can also be used in ornamental items. High strength, high elasticity, excellent wear and corrosion resistance etc., have made BMGs' suitable to be used in biomedical prostheses, load bearing biomedical implant applications, etc. High flexibility, high resistant to corrosion in the physiological environment and enhanced compliance to blood vessel biomechanics has made it suitable to be used as bone implants and replacements, stents and so on (Chen, 2011; Wang et al., 2004).

The physical properties like thermal diffusivity and the cutting conditions, such as cutting speed, depth of cut, etc. influence the shear flow mechanism and micro-structural changes in the material. BMGs have low thermal conductivity (~4W/m-K). Due to low thermal conductivity, plastic deformation energy that is converted to thermal energy in the cutting zone does not get a preferential exit into the machined workpiece. Rather, the majority of the thermal energy gets contained in the chips resulting in increased cutting zone-chip temperature, and light emission from the cutting zone. This phenomenon happens readily in macro scale cutting for larger cutting speed and depth of cut. However, in micro scale cutting as the chip load and the consequent mass transfer across the cutting zone are low, the thermal energy that is evolved in the cutting zone increasingly passes to the surrounding BMG material. This phenomenon increases the possibility of subjecting the surrounding material to have a thermal cycle of initially having increased temperature and subsequently cooling down. This effect causes the development of nano-scale crystallites in the amorphous matrix. The dispersion of nano-scale crystallites increases elastic modulus, tensile strength, etc. In some BMGs, plastic strain to failure also gets increased with increased volume fraction of nano-scale crystallites (Chen, 2011; Stephenson and Agapiou, 2016; Wang et al., 2004). Apart from this aspect, many other aspects prevalent in micro cutting such as minimum uncut chip thickness (MUCT) effect, cutting edge radius, ploughing, work material property and micro structural effect, thermal softening, chip formation, material removal mechanism and process mechanics, influence the machining responses (Ray and Puri, 2023).

Near net shape BMG micro-components can be made by forming processes, such as high pressure die casting processes. Forward extrusion, closed die forging, backward extrusion at super cooled state, hot embossing are some of the processes which are used for making BMG micro-parts. Micro-features may be made on BMG material that is used to create high temperature polymer insert by hot embossing. This insert is further used in injection molding for creating micro parts. Amorphous micro-structure, good process ability and large super-cooled liquid region of bulk amorphous metals make them highly promising candidates for thermoplastic processing (Kumar et al., 2009; Pan et al., 2008; Schroers, 2009; Wert et al., 2009). Thermoplastic forming (TPF) has emerged as a method for manufacturing of net shape metallic glass components. However micro-forming methods based on molding, possess some limitations in use due to the difficulty in creating deep cavity shapes and in de-molding process (Duan et al., 2007; Schroers, 2009). In this regard, micro-machining technology provides a reliable opportunity to shape BMG micro-parts. Mechanical micro-machining has more benefits than other alternative techniques, as it is capable of shaping complex three-dimensional micro-parts in a reliable and controllable way. Micro cut features machined on bulk metallic glass samples are shown in Figure 1. Micro-components often integrate multiple functions in one product. There is a challenging need to create complex surface features on these components with high dimensional and form accuracy, high surface and edge quality. Micro turning, micro drilling and most importantly micro milling are some of the notable micro-machining methods. These methods are indispensable in manufacturing micro- and miniature products using a variety of engineering materials (Ahmed and Jackson, 2014). For a better understanding and utilising micro-cutting for shaping BMG parts, this chapter discusses critical aspects of micro cutting mechanism, chip formation in cutting bulk metallic glass and several process performance aspects in micro-cutting of bulk metallic glass.

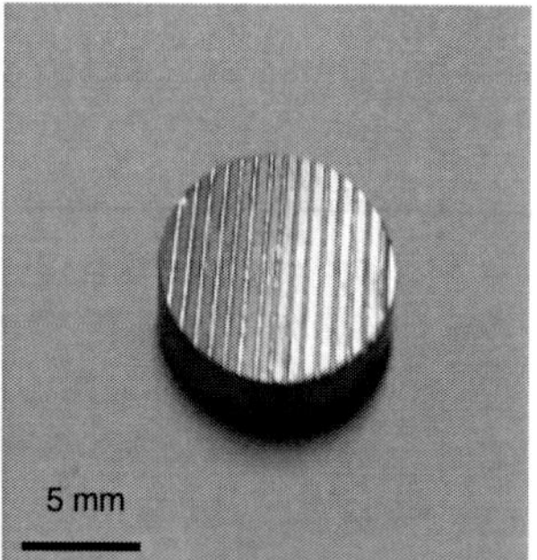

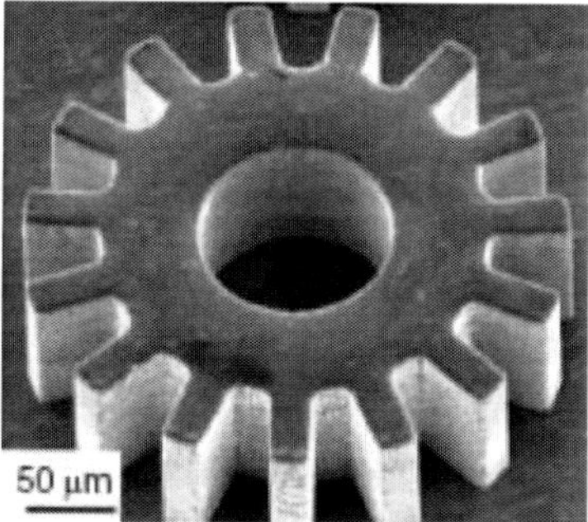

Figure 1. Micro cut features in Bulk Metallic Glass.

Fundamental Aspects in the Micro Cutting Process

Process Mechanics

Mechanical micro-machining with micro tools also referred as micro-cutting creates machined features by tool-work interaction at every point on the generated surface. Here also, as in macro cutting, the cutting edge element of the cutting tool separates work material as a chip from the parent workpiece by tearing and/or fracture. Micro-cutting processes are deterministic mechanical micro-scale finishing processes that can specify the profile of a three-dimensional micro-component by a well-defined tool path and surface generation. Typically, these processes include micro turning, micro drilling and micro milling. Mechanical micro-machining differs from conventional macro-machining due to reduced dimensions in the respective size of the machined features, cutting tool and cutting parameters. Material removal largely takes place in front of the rounded cutting edge and several aspects related to chip removal mechanism and process mechanics need to be considered in micro cutting. Micro cutting creates feature sizes ranging from tens of micrometres to a few millimetres. Commonly in micro features, at least in two dimensions the size is below 1 mm. Alternative definitions of micro-cutting are related with the dimension of the cutting tool or the edge radius of the cutting tool. The diameter of customized or commercially available micro end milling tools usually ranges from 10 μm to 1 mm having one or more cutting edges. The cutting edge radius normally lies within 5-6 μm. For precision and accuracy of the shape and size of the micro-features, material removal in micro cutting is done in smaller units with small chip cross sections and length of cut. The uncut chip thickness is comparable in size with the cutter edge radius and the cutting tool operates with a small depth of cut and a low chip load on the cutting edge (Chae et al., 2006; Dornfield et al., 2006; Huo and Cheng, 2013). The limits of the material removal with down-sized tool-work interaction in micro cutting are dictated by several factors, such as, minimum feasible uncut chip thickness, cutting edge radius, work material properties and its micro-structure, cutting tool dynamics, and so on. However, smaller material removal at the atomistic level is not achievable by mechanical micro-machining processes.

The chip formation process becomes increasingly difficult in micro cutting when the uncut thickness is low enough and is below a critical value. This limiting or the minimum feasible uncut chip thickness is referred as the minimum uncut chip thickness (h_m). Several research studies reported in the literature have inferred that minimum uncut chip thickness is dependent on the cutting edge radius and the properties of the work material. Considering a wide range of metallic materials, crystalline to amorphous, the ratio of minimum uncut chip thickness to the cutting edge radius was observed to vary from 0.1 to 0.5 (approx.). As the uncut chip thickness is comparable to the cutter edge radius, the ratio between the uncut material thickness and the cutting edge radius (h/r_e) plays a significant role in micro cutting (Huo and Cheng, 2013; Ikawa et al., 1992; Kim and Kim, 1995; Kim et al., 2004; Liu et al., 2004). Figure 2 schematically illustrates the influence of the uncut material thickness and the cutting tool edge radius on the cutting process, where the cutting tool either removes work material as a chip or ploughs and rubs the work material. When the uncut chip thickness (h) is larger than the minimum uncut chip thickness (h_m) level, material separation takes place and the shearing mechanism plays a significant role in the chip formation. When the uncut chip thickness level is around the minimum chip thickness (h_m) level, chip formation hardly takes place and the material is ploughed and pushed below the cutting edge. The ploughed material that flows below the flank face of the cutting tool undergoes elastic-plastic deformation with probable elastic recovery. Ploughing occurs due to the radial pressure applied on the work material by the rounded cutting edge. Material removal in micro-cutting may be visualized as a cutting phenomenon where the upper layer (of the total uncut layer) of the work material above a critical thickness rises up the rounded edge to form the chip, whereas the bottom layer of the incoming material gets ploughed by the rounded cutting edge (Albrecht, 1960; Connolly and Rubenstein, 1968; Sarwar and Thompson, 1982). Chip formation in micro cutting at a low feed rate has simultaneous involvement of shearing and ploughing mechanisms where the material above the depth of the minimum uncut chip thickness forms a chip and the material below the minimum uncut chip thickness level is ploughed and deformed under the cutting edge followed by some partial elastic recovery. For much higher uncut thicknesses, cutting takes place primarily by shearing; ploughing becomes less dominant and elastic recovery at the flank face may be considered negligible. The proportion of ploughing and rubbing forces in the overall cutting action of the micro cutting process is not insignificant. A larger proportion of the cutting energy is utilized in ploughing without contributing to the material removal and giving rise to the size effect (Inamura et al., 1993; Lucca et al., 1993; Waldorf et al., 1999; Wu et al., 2016). Various effects related to material removal and material flow around the cutting edge arising from the different conditions of uncut chip thickness and cutting edge radius are presented in Table 1.

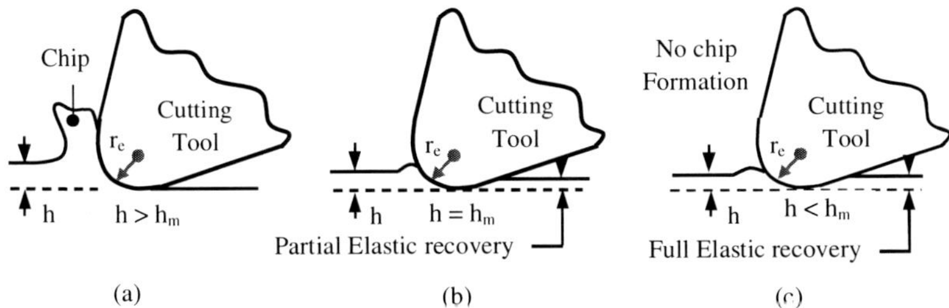

Figure 2. Influence of cutting tool edge radius on cutting mechanism (a) Shearing (b) Ploughing (c) Rubbing for different h/ r_e ratios and values of h compared to h_m.

Table 1. Influence of various conditions of uncut chip thickness (h) and cutting edge radius (r_e) on cutting.

Decreasing Ratio of (h/r_e) (r_{e3}> r_{e2}> r_{e1}) ↓		Decreasing uncut chip thickness (h_1> h_2> h_3) →		
		h_1 (> h_m)	h_2 (= h_m)	h_3 (< h_m)
	h/r_{e1}	Cutting	Ploughing	Rubbing
	h/r_{e2}	Cutting	Ploughing	-
	h/r_{e3}	Cutting	-	-

The uncut chip thickness in macro-scale cutting is usually much greater than the cutting tool edge radius. For example, in macro milling with a 5 mm diameter end mill, the uncut chip thickness can be 0.3 mm whereas the cutter edge radius can be as small as 0.003 mm. Similarly in macro-scale turning the uncut chip thickness can be 0.5 mm whereas the cutting edge radius of the single point cutting tool can be 0.005 mm. Therefore in macro cutting, for a cutting tool with a positive nominal rake angle, the effective rake angle at the average uncut layer thickness is almost equal to the nominal rake angle. In this case, the rounded segment of the cutting edge can be neglected and the straight rake face confronts the majority of the uncut layer. In contrary to macro-scale cutting, a different scenario exists in micro-cutting. In a typical micro-milling process with a 0.3 mm diameter micro end mill, the maximum uncut chip thickness can be as small as 0.001 mm where the cutter edge radius can be 0.003 mm. Likewise, in micro-scale turning, the maximum uncut chip thickness can be 0.001 mm whereas the cutting edge radius for polycrystalline diamond or cubic boron nitride cutting tool can be 0.002 mm. For a single crystal diamond tool the cutting edge is even sharper with a cutting edge radius around a few hundred of nanometres. The cutting tool in micro cutting process may have a positive nominal rake angle whereas the effective rake angle at the average uncut layer thickness is largely negative. As shown in Figure 3, when the uncut chip thickness 'h' exceeds 'h^*' (maximum uncut layer confronted by the rounded cutting edge), the cutting scenario is shear dominant and the approaching uncut material confronts the tool face of the cutting tool with edge radius 'r_e' and nominal rake angle 'γ_n'. 'h^*' mentioned here is equal to [$r_e(1 + \sin\gamma_n)$]. When 'h' is less than 'h^*', the incoming uncut material confronts the rounded edge of the cutting tool at effective rake angle 'γ_e' which is negative corresponding to the cutting depth 'h', and is equal to [$-\sin^{-1}\left(1 - \frac{h}{r_e}\right)$].

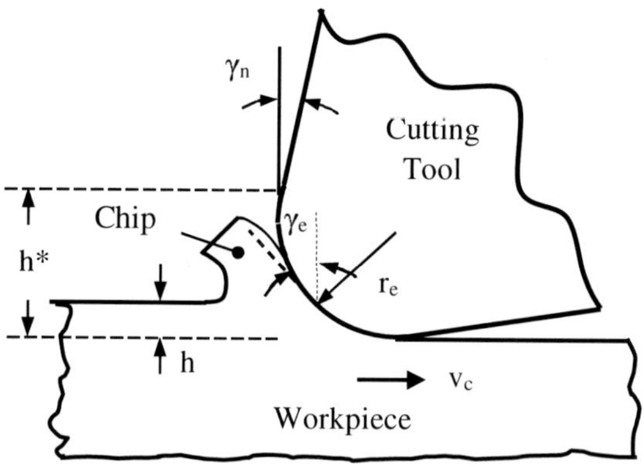

Figure 3. Schematic of the effective rake angle (γ_e) in micro scale cutting.

The incoming work material flow in the ploughing dominant cutting scenario is restricted due to the barrier of the highly negative rake angle effective at the cutting point. The approaching work material towards the rounded cutting edge bifurcates at some separation point in this zone of restriction ahead of the cutting tool edge. This separation point may be described as a stagnation point having a location angle α_s as shown in Figure 4. The inflowing work material above the level of this separation point joins to the flowing out chip whereas the lower layer below the separation point is pushed below the cutting edge to become part of the machined work material. The development of the separation zone may also be viewed to be due to locking up of the approaching work material towards the cutting edge. The locked up work material hardly finds a way to pass away with the chip and forms a very small 'build-up' adhering in front of the cutting edge as a stable dead metal zone (Abdelmoneim and Scrutton, 1974).

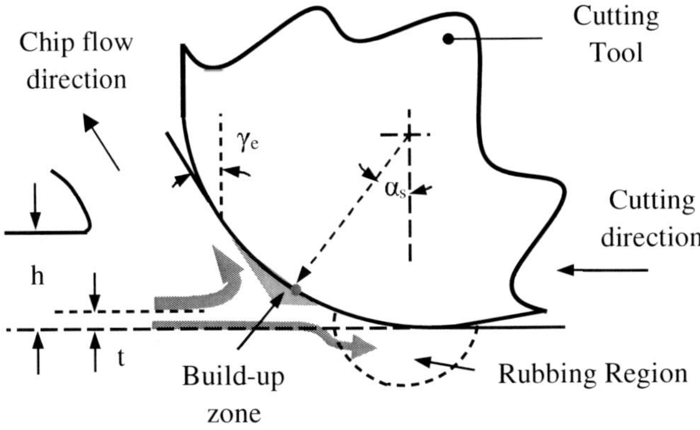

Figure 4. Flow of work material around rounded cutting edge [α_s ploughing angle].

As the cutting tool advances relative to the work material with the uncut chip thickness much greater than the cutting edge radius in macro-scale cutting, the material ahead of the cutting tool is severely compressed subjecting the work material to shear through the shear zone along slip planes with highest defect densities. The shearing is accomplished with the failure lines moving along the grain boundaries for polycrystalline materials. The material is separated as an aggregate of grains by plastic deformation typically observed in ductile crystalline substances. However, in the case of brittle substances, the failure is similarly initiated at the grain boundaries due to brittle fracture with cracks propagating rapidly across the material. When the uncut chip thickness gradually decreases, it becomes comparable in size with the average grain size and therefore, removal of the material takes place across the individual grains. In this case, the typical failure path moves across the grain instead of along the grain boundary. Defects such as dislocations that help in slip play a significant role in facilitating the failure path. Obviously as the uncut layer thickness reduces from macro to micro level, the density of defects becomes lesser making the resistance to the material separation to increase significantly. This is the reason for increased strengthening effect of the work material at low depths of cut in micro cutting. For amorphous substances such as bulk metallic glasses, deformation during cutting is associated with the development of shear bands. The deforming volume is substantially reduced in micro cutting causing increased strengthening in bulk metallic glasses (BMGs), as the rapid propagation of shear bands is suppressed due to a lack of preferential sites from where the embryonic shear bands originate (Jang and Greer, 2010; Kuzmin et al., 2012; Schroers and Johnson, 2004). Work material strengthening and ploughing in micro cutting results in nonlinear increase in the specific cutting energy or specific cutting force. This phenomenon is described as the 'size effect' (Liu and Melkote, 2005; Malekian et al., 2009a; Ray et al., 2020a).

Chip formation in micro-cutting is a highly dynamic process. The dynamic chip thickness is highly non-uniform due to the process dynamics, tool run-out, tool wear, vibrations, regenerative effects, etc. Chip formation in micro-cutting experiences changes in the shear angle of the slip planes in the cutting zone and variation in the chip thickness. Depending upon the material properties and the process conditions, wavy, segmented, discontinuous, elemental, and particle chips are formed in micro-cutting. In orthogonal shaping or turning processes, the uncut chip thickness may be considered uniform provided the vibrational and regenerative effects are insignificant. However, in micro milling process, the uncut chip thickness is not uniform as a cutting edge cuts material during its rotation. Micro-channels are common micro-features that are often produced by full immersion micro-milling process where the micro-cutter diameter determines the width of the micro-channel and the axial depth of cut determines the micro-channel depth. For a two fluted cutter in full immersion micro milling (Figure 5), when a cutting edge completes its rotation in the cut, the subsequent cutting edge encounters the material left from the earlier cutting teeth pass. When the projected feed per tooth at the cutting point (along the radial distance of the cutting edge from the centre of the tool axis), is less than the minimum uncut chip thickness, chips are generated once in several rotational passes till a sufficient material thickness accumulates ahead of the cutting edge. A cutting edge that is removing material experiences a larger uncut chip thickness due to accumulated left-out uncut material by the previous cutting edges. Thus, the initial point of chip formation may start from a smaller immersion angle than the earlier pass, and the actual material cutting region is extended for the cutting edge that is removing material if previous cutting edges did not cut any material (Filiz et al., 2007). This aspect has been illustrated in Figure 6. The analysis of chip volumes in a study of micro end milling of brass at different feed rates applicable in micro-cutting revealed that the ratio of measured chip volume to nominal chip volume approached the value of unity at higher feed rates, whereas, at lower feed rates, the ratio was much higher indicating that chips were not formed in each pass of a cutting tooth. In this study, examination of feed marks revealed that lower feed rates resulted in several non-cutting passes (Kim et al., 2002).

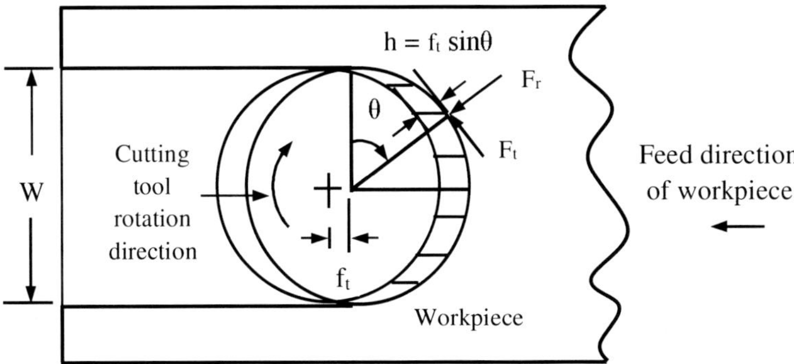

Figure 5. Schematic of full immersion micro milling process. [h (instantaneous chip thickness) = $f_t\sin\theta$ (approximate projected feed per tooth at the cutting point); θ = immersion angle at the cutting point; f_t = feed per tooth; F_t, F_r = tangential and radial forces at the cutting point; W = width of cut].

In addition to this, the tool run-out effect that arises due to improper tool symmetry, imperfect tool alignment in the tool holder, mismatch between the tool holder and spindle, positional error in spindle bearings, etc. results in the variation of the chip load on the cutting edges leading to significant force variation, rougher surface generation, unequal tool edge wear, etc. Micro-cutting performance in terms of accuracy and surface finish largely depends upon the dynamic performance and robustness of the machine tool, its allied tools and accessories. The key components of a micro machining system include machine structure, spindle, positioning stages and control system, fixtures and clamping devices. Studies on micro-cutting are mainly carried out in commercially available high cost precision machining centres or in developed specialized miniature machine tools. However, in both type of systems, integration of proper sub-systems to obtain an improved configuration of the machine tool with high rigidity, good positional accuracy, precision motion control with good repeatability, adequate dynamic stability and adequate isolation from the environmental effects, vibration control, etc., ensures good performance in cutting (Bang et al., 2004; Dornfield et al., 2006; Weck et al., 2001).

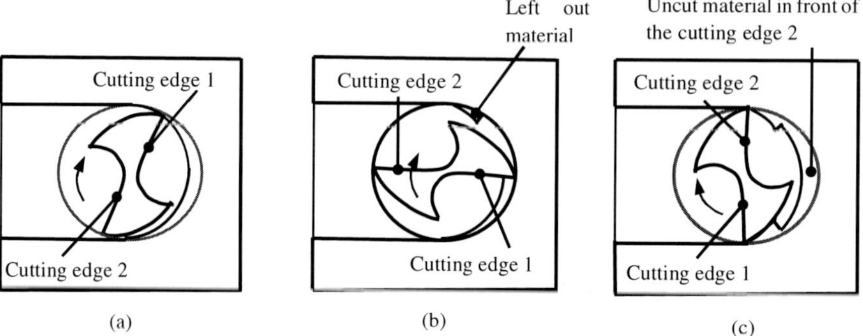

Figure 6. (a)-(c) Excess material accumulates ahead of cutting edge 2 due to work material left uncut by cutting edge 1.

Chip Formation Mechanism in Bulk Metallic Glass

Analysis of chip formation and study of chip morphology reveals the underlying deformation and fracture mechanisms in machining processes. Understanding the fundamental mechanisms of chip formation helps in understanding the influence of the process variables on the micro cutting process and is critical to optimise the machining performance and development of improved micro-cutting tools. Limited studies have been done on the analysis of chip formation in cutting bulk metallic glass material with some of the studies particularly at low cutting speeds and at depths of cuts higher than the range commonly used for micro cutting. However these studies have guiding importance to understand the mechanism in micro-scale cutting. Most of these studies have reported segmented chip formation where the chip emerges in segments separated by a narrow thin zone referred as primary shear zone (PSZ) of approximately 0.01-0.03 μm thickness (Dhale et al., 2019; Ding et al., 2020; Jiang and Dai, 2009; Maroju and Jin, 2019; Ueda and Manabe, 1992).

The serrated form of the chip occurs due to the relative sliding of the adjacent segments. As the material starts to separate out from the workpiece to form a chip, the deformation is accommodated with the emergence of multiple shear bands. The segments have a zone comprising of secondary shear bands (SSBs) and a narrow zone comprising of closely packed repeated shear bands called primary shear zone (PSZ). Segmentation in BMGs occurs along PSZ. The segmentation is schematically shown in Figure 7. The formation of secondary shear bands within a segment acts as a mechanism to provide plasticity character to the deforming material and prevents material to fracture rapidly along a plane (Ding et al., 2020). Material deformation involving SSBs in cutting BMGs is inhomogeneous in nature and is in the form of localised shear bands due to free volume generation. Several crystalline materials that result in segmented chip formation have periodic adiabatic shearing due to thermal instability. The plastic deformation of crystalline materials occurs due to dislocation movement and slip within the grains and/or along the grain boundaries. In BMGs, dislocation mediated slip is not observed. Rather, BMGs under deforming load undergo shearing along shear bands governed by localised shear stress driven free volume instability. As mentioned earlier, BMGs have amorphous atomic structure without having any long range atomic order. Free volume in materials such as BMGs is a void space among a cluster of atoms. Due to intense stress concentration at the cutting tool tip, the balance is broken between generation and dissipation of free volume in the deforming volume. There is continuous generation as well as dissipation of free volumes in the cutting of BMG due to the rearrangement of the atoms during deformation. This results in high free volume generation or concentration. Excess free volume generation or free volume concentration in a highly complex dynamic process such as chip formation, leads to the development of nano/micro voids within the deforming material. Atomic rearrangements due to the deforming energy also increase the temperature of the deforming material. These nano/micro voids enlarge, distort and coalesce, and in a complex state of stress, develop micro-cracks. This is schematically represented in Figure 8. Continuous propagation of micro-cracks and increased temperature cause chip root to break with some viscous slipping in narrow shear transformation zones, eventually separating the segment by considerable slippage from the adjacent or newly developing segment. The yielding of material in PSZ is due to the unstable propagation of huge quantity shear transformation zones (STZs) in PSZ (Maroju and Jin, 2019). Analysis of the interface of the chip and the workpiece (may be referred as chip root surface) in studies revealed that material around the micro-cracks evolves into dimple patterns. The smeared material in the high temperature viscous shearing regime around the chip root reveals a dimple appearance in micro scale. The appearance of micro scale dimples on the sheared surface resembles ductile tearing along with crack propagation during segmentation. Chip formation in BMG cutting as described by some researchers is through ductile-brittle fracture mode (Dhale et al., 2019). As cutting continues, a segment slides over the segment behind it along the PSZ and also moves up the tool rake face. The BMG material within the narrow PSZ is generally subjected to higher temperatures than its melting point. Friction at the cutting tool-chip interface also causes temperature of the chip material to increase. The increase in temperature of the material and its subsequent cooling creates a thermal cycle and nano-crystals nucleate within the deforming material. The nano crystals block catastrophic shear failure, causing branching and bifurcations of the shear bands.

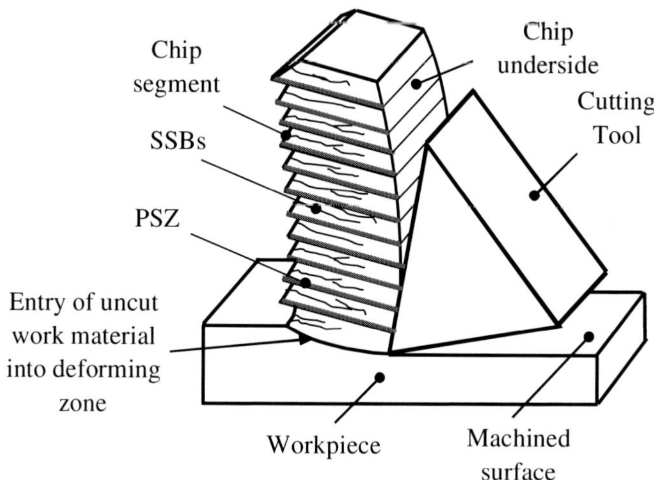

Figure 7. Schematic of chip segmentation in BMG cutting.

To understand the chip formation mechanism in orthogonal micro-cutting of Fe-based amorphous alloy, a micro-cutting set-up was used inside the scanning electron microscope in a study (Ueda and Manabe, 1992). The study observed chip formation with lamellar structure due to the periodic formation of the localized shear bands. The chip formation process was analysed by simulation considering adiabatic deformation in the primary deformation zone. The simulation results regarding lamellar spacing, shear band angle, etc. were found to have agreed well with the experimental results. In a study on the chip formation in Zr-based BMG ($Zr_{41.2}Ti_{13.8}Cu_{12.5}Ni_{10.0}Be_{22.5}$) lamellar structure chips were observed (Jiang and Dai, 2009). Lamellar structure occurred due to repeated shear band formation originating at the tool contact surface, propagating toward the free surface in the primary shear zone. These shear bands were periodically located with constant spacing. However, no characteristic fracture patterns such as vein type patterns were observed in the chips. Chip morphology revealed a close resemblance of lamellar type chips with serrated chips that are formed in machining crystalline alloys. Observation through scanning electron microscope (SEM) revealed that the shear band spacing and shear displacement were smaller with the characteristic smoother free surface for the BMG material. Several secondary shear bands were observed along with the primary shear band in the PSZ due to the shear localization at very small cutting speeds. A thermo-mechanical orthogonal cutting model, considering force, free volume and energy balance in the primary shear zone was developed in this study to characterise lamellar chip formation. Theoretical simulation along with the experimental observations in the study provided a clearer physical picture of the lamellar chip formation mechanism in cutting bulk metallic glass.

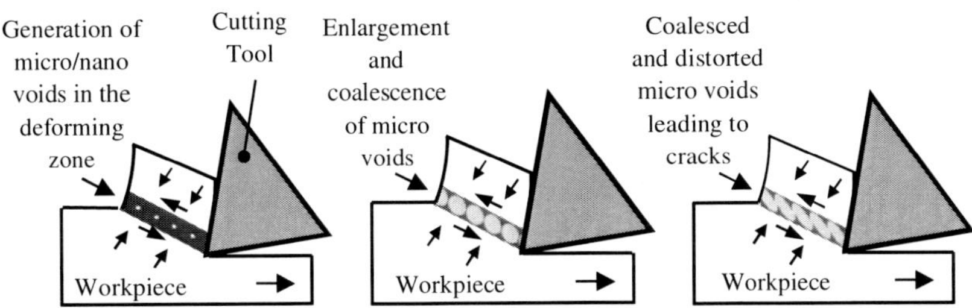

Figure 8. Schematic of the process from development of micro/nano voids to further enlargement, coalescence and distortion leading to micro-racks in the deforming zone.

Analysis of Process Performance Measures in Micro-Cutting of BMG

Quality Characteristics of Micro-Machined Features

Micro-component design and fabrication have acquired an enormous scientific interest in various application areas. The quality of these micro components is of major concern due to their stringent functional requirements. Surface and edge quality, profile, form and dimensional accuracy of these micro components represent the vital quality aspects. There is a growing need to develop suitable technology that is capable of creating intricate and precise micro features. Micro-cutting processes are preferably used to produce different shaped micro cut features. These processes include micro turning, micro drilling, and most importantly micro milling. These processes use diamond cutting tools, polycrystalline cubic boron nitride (PCBN) tools, coated and uncoated carbide tools.

Analysis of Surface Roughness in Precision Turning/ Micro-Turning/ Micro-Milling of BMG

Diamond turning is a precision cutting process that uses a sharp diamond cutting tool. With regard to the work piece surface quality, surface roughness in the nanometric level and surface waviness in the submicron level may be achieved in diamond turning. Diamond cutting tools with fine cutting edges in tens of nanometers are used to achieve such a low level of surface roughness (Balasubramaniam and Suri, 2011). For producing axis-symmetric micro-features on bulk metallic glass samples, diamond turning is an effective process. Micro-cutting of bulk metallic glass with a diamond cutting tool involves a complex interaction between the diamond cutting bits and bulk metallic glass material. The material removal is influenced by several factors such as work-tool material properties, cutting tool geometry, use of coolant, cutting parameters such as depth of cut, feed rate, cutting velocity and so on. Literature had reported some studies on the characteristics of machined surfaces, generated in the precision turning of bulk metallic glass using different inserts including diamond cutting tools. A study on the cutting characteristics of bulk metallic glass ($Zr_{41.2}Ti_{13.8}Cu_{12.5}Ni_{10.0}Be_{22.5}$) showed that diamond turning of a BMG cylindrical sample (Bakkal et al., 2004) resulted in a better finish than aluminium alloy Al6061 and stainless steel SS304 at depth of cut and feed rate of 0.5 mm and 50 μm per revolution respectively in the cutting speed range of 0.38-1.52 m/s. In this study, roughness attained in BMG with the polycrystalline diamond tool was 0.62 μm whereas roughness attained in Al6061 and SS304 were 1.07 μm and 1.5 μm respectively. However, this study showed that a much lesser roughness of 0.35 μm could be attained in the BMG sample with the WC-CVD tool at a cutting speed of 0.38 m/s. A similar type of turning experiments were conducted in another study (Fujita et al., 2005) with BMG ($Zr_{65}Cu_{15}Ni_{10}Al_{10}$ and $Pd_{40}Cu_{30}Ni_{10}P_{20}$), steel (JIS SGD-400D) and brass (JIS H C3604) samples at feed rate of 50 μm, and depth of cut of 0.05 mm with cutting speed in the range of 0.08-2.5 m/s. The study also indicated that lower roughness values were attained in BMG than in other crystalline alloys. In this study, the minimum roughness value obtained with BMG work material was 0.08 μm which was lower than that obtained from many finishing level precise cuts. This indicates that a precise level finish is easily obtained in BMGs.

Diamond turning of Zr-based BMG ($Zr_{41.25}Ti_{13.75}Ni_{10}Cu_{12.5}Be_{22.5}$) using a single crystal diamond cutting tool with a nose radius of 0.25 mm under the application of atomized coolant showed that cutting parameters such as depth of cut, feed rate and spindle speed significantly influenced the surface morphology of the generated surfaces (Han et al., 2015). The range in depth of cut, feed rate and cutting speed in this study were 1-3 μm, 5-20 μm per revolution and 1.83-7.32 m/s respectively. SEM images taken in the study revealed that grooves were present on the surfaces at higher feed rates and depth of cuts. The grooves were in-homogenous when cutting was carried out at a depth of cut of 1 μm. The average linear surface roughness values obtained were 3.39 nm and 7.47 nm at depth of cut of 1 μm and 3 μm respectively. An increase in the feed rate from 5μm/rev to 10 μm/rev at cutting speed 7.32 m/s and depth of cut 3 μm apparently smoothened the surface. Three-dimensional morphologies of the turned surfaces of the BMG sample revealed the interaction between the cutting tool and the work piece. Raising the feed rate from 5 μm to 10 μm increased the width of grooves developed on the surface from 5 to 10 μm. However, the depth of the grooves was shallower for a feed rate of 10 μm than at 5 μm. This might be due to the adiabatic heating developed during the separation of the chips as opined by the researchers of the study. As the thermal conductivity of the Zr-based BMG is typically low (~4 W/m-K), the adiabatic heating developed during the formation of the chips would have raised the localized temperature in the cutting zone. The heat transfer rate to the viscous softened layer of the cutting zone at higher feed rates was larger than that at the low feed rate, resulting in an increased volume of the softened layer that resulted in generating irregular patterns with smaller groove depths. The study revealed that a larger depth of cut contributed significantly to the formation of sinusoidal-like grooves. Furthermore, the increased feed rate resulted in a smaller roughness on the machined surface but compromised the process stability causing irregular grooves to appear on the surface. The study further reported that surface morphology was substantially influenced by the cutting speed where high cutting speeds generated low roughness in BMG samples.

A study on the precision turning of Zr-based bulk metallic glass ($Zr_{52.5}Ti_5Cu_{17.9}Ni_{14.6}Al_{10}$) using diamond and PCBN cutter provided further understanding of the micro machinability of bulk metallic glass regarding the characteristics of the generated surface (Chen et al., 2017). This study obtained the ratio of the critical uncut chip thickness in BMG to the diamond tool edge radius to be 0.65 in the range of cutting speed, feed rate and depth of cut of 0.84-1.68 m/s, 0.5-10 μm, and 0.6-20 μm respectively. The ratio of uncut chip thickness to the cutting edge radius obtained in this study is higher than obtained in most crystalline metallic materials (less than 0.5). This might be due to the high elastic limit of bulk metallic glass (~2%). The critical uncut chip thickness refers to the limiting uncut layer of work material below which chip formation ceases to occur. The study showed a decrease in surface roughness with a decrease in feed rate up to a certain value and then with a further decrease in feed rate (in the very low feed rate range) the roughness value showed an increasing trend. This phenomenon may be attributed to the excessive ploughing at very low feed rates and side flow of the work material that might have accumulated over the generated surface. The variation of the surface roughness (R_a) against the variation of the cutting speed at a feed rate of 5 μm and depth of cut of 10 μm showed an increase in surface roughness with the increase in cutting speed in the range of 0.84-1.47 m/s. However, the roughness values decreased as the cutting speed was increased beyond 1.47 m/s in the range of 1.47-1.68 m/s.

Variations of surface roughness with feed rate and cutting speed are represented in Figure 9. This variation of roughness may be attributed to the adverse effects of the built up zone locked ahead of the cutting edge in precision cutting and also its disappearance as the temperature in the primary cutting zone and the cutting tool-chip interface increased with the increase in the cutting speed. The surface topography of the precision turned work samples analysed in this study revealed a viscous flow of the work material in the micro-cutting process.

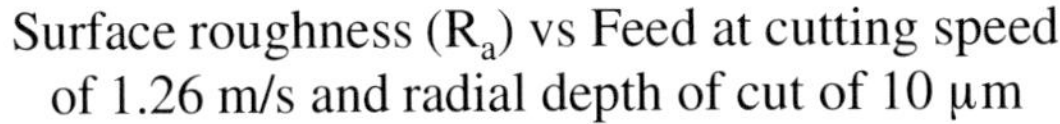

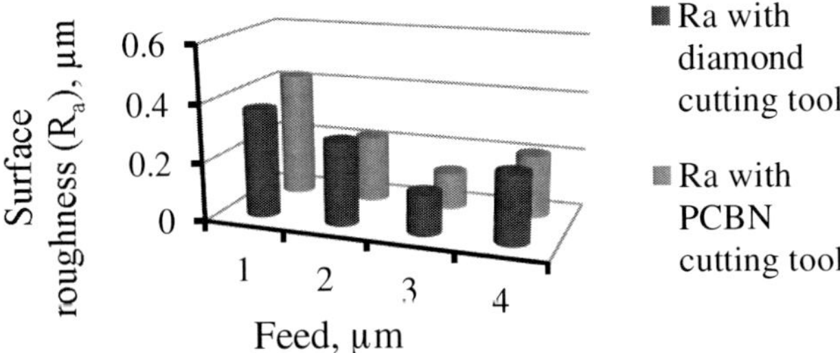

Surface roughness (R_a) vs cutting speed at feed of 5 μm/rev. and radial depth of cut of 10 μm

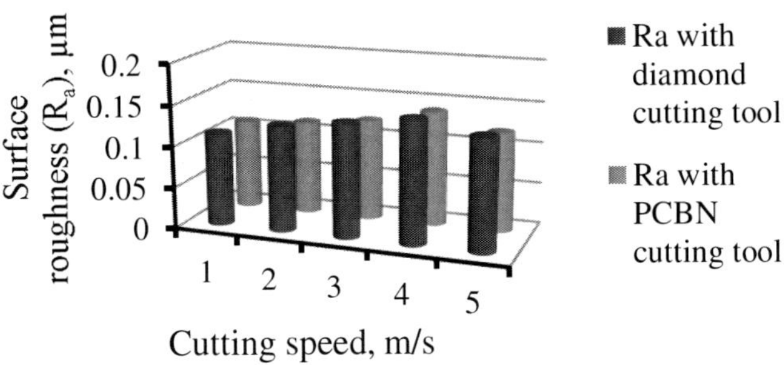

Figure 9.Variation of average surface roughness in BMGs with diamond and PCBN cutting tools at different cutting conditions.

An analysis of the influence of the cutting parameters on the generation of surface roughness at the floor of the micro channels fabricated in BMG ($Zr_{68}Cu_{12}Ni_{9}Ti_{11}$) sample by full immersion micro milling process was reported in a study (Ray et al., 2020b) that used 0.3 mm diameter micro-end mill having average edge radius of 3 μm. The experimental process was carried out with cutting parameters such as feed per tooth (f_t), axial depth of cut (a_p) and cutting velocity (v_c) in the range of 3.3–6.7 μm, 11.6–28.4 μm and 0.036- 0.090 m/s (spindle speed (N) 2318-5682 rpm) respectively. The variation of the average area roughness (S_a) obtained in the study is depicted in Figure 10.

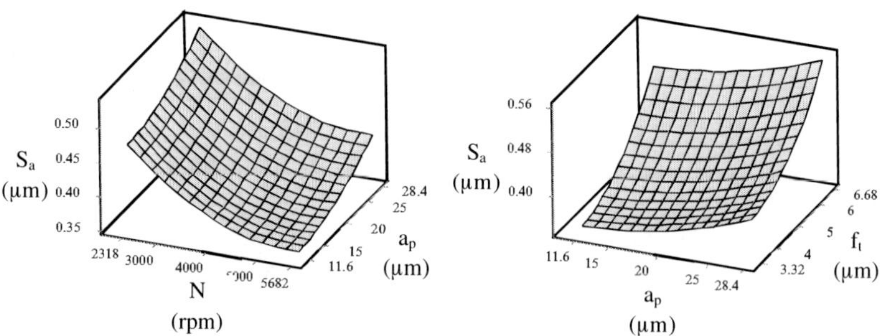

Figure 10. (a) Surface plot of S_a with N and a_p at $f_t = 4$ µm. (b) Surface plot of S_a with a_p and f_t at N (spindle speed) = 4000 rpm. Spindle speed of 4000 rpm corresponds to cutting speed (v_c) of 0.063 m/s.

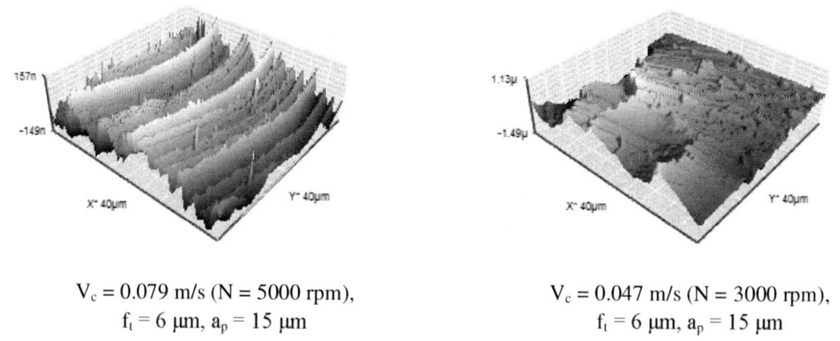

$V_c = 0.079$ m/s (N = 5000 rpm), $f_t = 6$ µm, $a_p = 15$ µm

$V_c = 0.047$ m/s (N = 3000 rpm), $f_t = 6$ µm, $a_p = 15$ µm

Figure 11.Three-dimensional micro end milled surface of the BMG sample at various cutting conditions using atomic force microscopy.

Area surface roughness (S_a) parameter provides comprehensive information on the surface texture characterization better than average linear roughness R_a. Three-dimensional roughness measurements with a sampling area of 40µm×40µm were taken at different locations along the length of the micro channels. The plots of surface roughness against the cutting parameters showed that surface roughness increased with the increase of feed per tooth when the feed rate was higher than the cutting edge radius. At very low values of feed per tooth, surface roughness showed an increasing tendency with the decrease of feed per tooth. In this study, surface generation of the micro-channel floor involved the secondary (end) cutting edge of the micro tool. The plot depicted in Figure 10 also showed that surface roughness decreased with the increase in the cutting speed. Three-dimensional profiles of the micro end milled surface of the BMG sample using atomic force microscopy at different levels of the cutting parameters are depicted in Figure 11. The effect of low feed rate in cutting action and the surface generation of the cutting edges of the micro end mill in full immersion micro end milling are depicted in the schematic Figure 12. The indenting pressure of the cutting tool on the work material might have compressed and plastically deformed the material to squeeze it out upward (due to peripheral/primary cutting edge) and/or sideward (due to bottom/secondary cutting edge). Some of the plastically deformed material that is ploughed aside at very low feed rates in micro milling might had stick to the newly machined surface resulting in elevated formations on the feed mark ridges (Figure 12). This deteriorated the machined surface quality and increased the generated surface roughness over the theoretical roughness obtained from tool geometry and the feed rate in cutting. Ploughing and material side flow influenced the actual roughness value at very low feed rates.

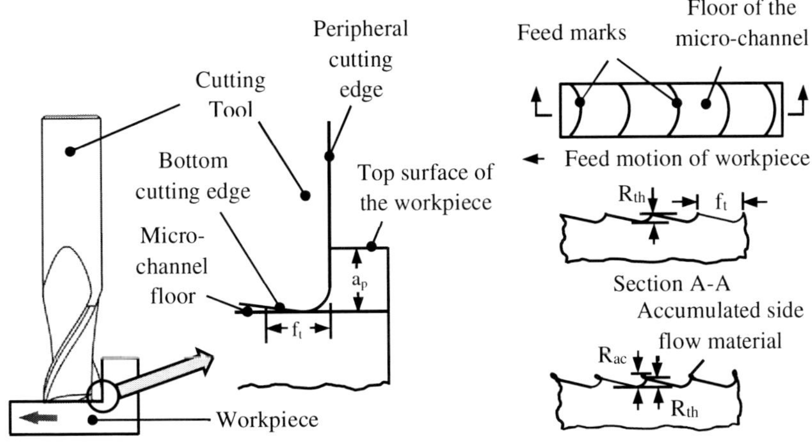

Figure 12. Influence of low feed rate in cutting action and the surface generation in micro end milling. [f_t = feed per tooth; a_p = axial depth of cut; R_{th} = theoretical surface roughness; R_{ac} = actual surface roughness].

Analysis of Burr Formation in Micro-Cutting of BMG

Surface and edge quality are important aspects in the quality of micro parts fabricated using precision machining technology. Burrs on the edges of these micro parts cause difficulty during its use. Burrs are undesirable thin protrusions of the work material that remain adhered to the machined edges. Burrs are formed due to several reasons such as tearing of a material, material bending and adhering to the edge of a machined profile at the end of cut (roll over burrs), material side flow due to intense plastic deformation (Poisson burrs), etc. Micro-cutting processes develop burrs of sizes comparable to the feature sizes and are difficult to remove due to the miniature shape of the machined features. Minimization of burr formation needs a proper understanding of the burr formation mechanism and an analysis of the influence of the machining parameters on the development of burrs. Though several micro parts are made of BMG material, however, limited studies were reported in the literature regarding the formation of burrs in micro cutting BMG.

Burr formation in drilling of Zr-based BMG ($Zr_{52.5}Ti_5Cu_{17.9}Ni_{14.6}Al_{10}$) at different feed rates (1.25-10 mm/min) and cutting speeds (0.1-1 m/s) using high speed steel and tungsten carbide drills (1 mm and 2 mm diameter) were analysed in a study (Bakkal et al., 2005) to understand the characterization of burr formation in drilling under varied cutting conditions. The size of the drill and the range of the feed rate used in the study were much higher than that normally used in micro cutting. Nevertheless, the results and the analysis have a guiding significance in understanding the issue of burr formation in BMG material. Burrs formed at the entry side edges were primarily Poisson burrs along with the remnants of the removed chips adhering to the edges of the drilled hole. The exit burrs mainly rollover type, were found to be larger than the entry burrs. HSS drills generated larger sized entry burrs than WC-Co drills. Relatively smaller sized burrs were seen on the drilled edges at higher feed rates. The higher cutting speed generated light emission and molten chips. Low feed rate (1.25 μm) and higher cutting speed (0.52 m/s) produced drilled holes where the edges were having melted material on the edges. This could be likely due to the high temperature generated at higher cutting speeds. Higher feed rate (10 mm/min) and cutting speed (1 m/s) generated plastically deformed blue coloured oxide covered burrs. Crown shaped exit burrs were identified in the study. A higher feed rate resulted in reduced burr size on the exit side. The experimental study noted that tool wear, tool run-out and ductility of the material also contributed significantly to the shape and development of burrs.

A study on micro drilling BMG (Zhu et al., 2013) with a feed rate in the range of 0.3-6.6 μm and cutting speed in the range of 0.21-0.74 m/s showed opposite results to that obtained in the aforementioned conventional drilling process where the exit size burrs were larger than the entry side burrs. This study on micro drilling obtained a larger burr height on the entry side than on the exit side. At the microscale, low feed rate, low ratio of uncut chip thickness to the cutting edge radius and highly negative effective rake angle resulted in significant ploughing and side flow of the plastically deformed work material particularly at the periphery during the entry of the micro drill into the workpiece. The Poisson burrs developed on the entry side were larger in size than the rollover burrs developed on the exit side. Burr formation at the top edges and exit burrs on the exit end edges of the channels made in Zr-based BMG sample were also analysed in a meso-end slot milling process (Bakkal and Naks[idot]ler, 2009) where a 1 mm diameter end mill was used at cutting speed of 0.26 m/s and in the feed rate range of 0.25-2 μm. Slots produced at feed rate of 0.25 μm were cleanest in appearance and the burrs developed at the edges increased with the increase in feed rate.

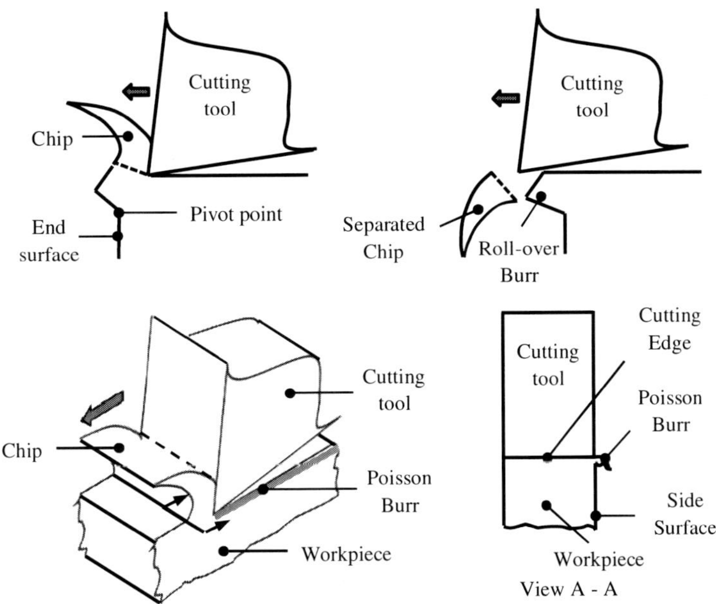

Figure 13. Schematic of roll-over and Poisson burr formation mechanism.

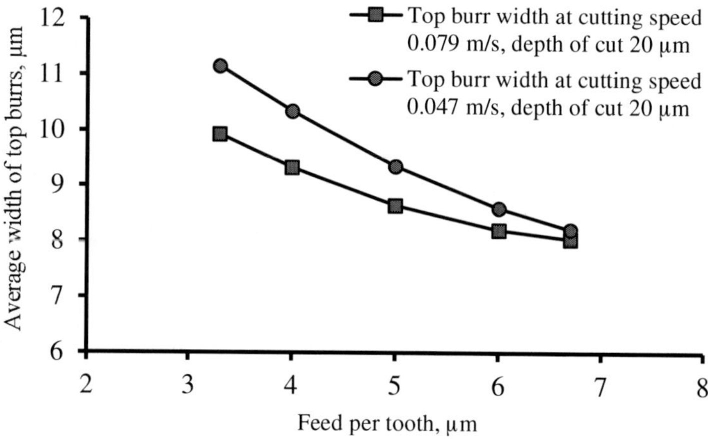

Figure 14. Variation of top burr width in micro end milled BMG (Zr_{68} Cu_{12} Ni_9 Ti_{11}) vs feed per tooth at various cutting conditions.

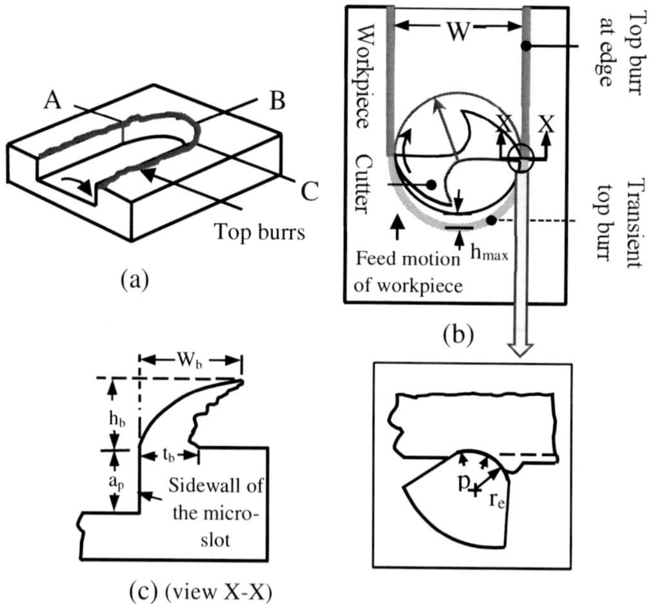

Figure 15. (a)-(b) Geometry of micro-channel cut by end milling process with two fluted cutter and location of top burrs; (b) Exaggerated view of burr formation zone; (c) Cross sectional geometry of top burr (view X-X). [W_b = width of top burr, h_b = height of top burr, a_p = axial depth of cut, t_b = burr root thickness].

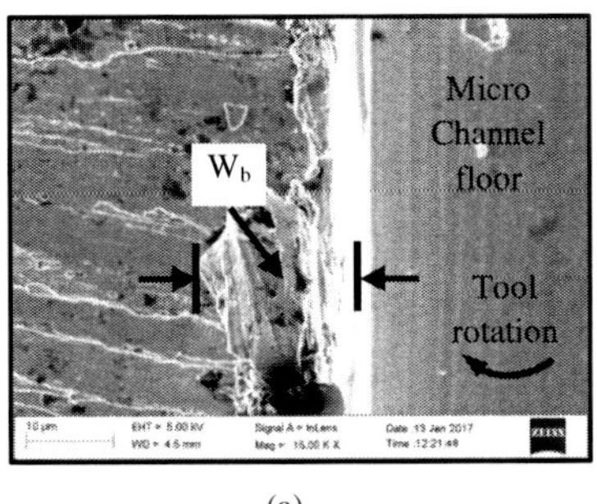

(a)

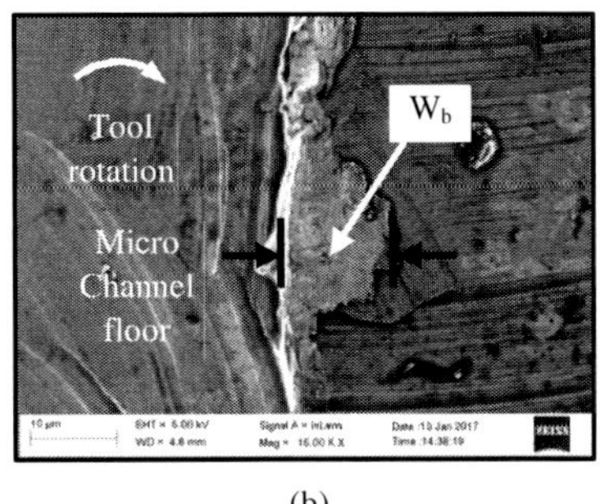

(b)

Figure 16. (a) Burr formation at up-milling side at cutting condition of $v_c = 0.047$ m/s, $f_t = 4$ μm, $a_p = 25$ μm (b) Burr formation at down-milling side at cutting condition of $v_c = 0.047$ m/s, $f_t = 5$ μm, $a_p = 20$ μm in full immersion micro-end milling. [W_b = Width of top burrs].

Micro-slotting is an important micro cutting process accomplished through micro end milling process to create micro channels. Two types of burrs are formed in the micro-slotting process. They are Poisson burrs and roll-over burrs. Roll-over burrs are formed at the entry and exit of the micro channels. The top Poisson burrs are formed at the top edges of the micro channels and occur over the length of the machined edge. The burr formation mechanism is schematically represented in Figure 13. Removal of the top burrs in micro-channels is difficult and the rectification process may affect the edge profile. The top burrs are formed when the sidewall material bulges out upward being subjected to ploughing and indenting pressure by the rounded cutting edge of the rotating end mill within a short but repeated engagement along the machined length. The amount of bulging and the burr size depends on the material properties, ploughing by the rounded cutting edge and the pressure developed due to the radial force applied by the cutting tool at the sidewall. The indenting pressure depends on the radial force which in turn depends on the cutting parameters. A suitable model of burr formation helps to estimate the extent of burr formation and quantify the dimension of the developed burrs. Limited modelling studies were reported that analysed the mechanism and characterization of different types of burrs formed in micro-milling (Lakkala et al., 2011; Ray et al., 2019a). In an experimental study of micro-milling bulk metallic glass (Ray et al, 2019b) that used 0.3 mm diameter uncoated micro-end mill of edge radius 3 μm, the width of the top burrs formed on the upper horizontal edges of the micro channels were measured for different cutting conditions. The experimental investigation was carried out with feed per tooth in the range of 3.3–6.7 μm, axial depth of cut in the range of 11.6-28.4 μm and cutting velocity in the range of 0.036-0.090 m/s. The variation in the average width of top burrs is depicted in Figure 14. The study indicated that a lower feed rate resulted in higher top burr size. When the cutter edge is at or near to point A, as shown in Figure 15, the inflowing material towards the cutter edge gets severely ploughed and plastic deformation of the material underneath the cutter edge (during its short duration stay in vicinity to the sidewall) caused the formation of top burr at the sidewall. The burr formed at point A gets adhered to the sidewall edge. At low feed per tooth, ploughing occurred for a longer time than for higher feed per tooth values. Thus the lateral bulging of material occurred more intensely at low feed rates. For higher feed rates, cutting action with chip formation starts earlier than for lower feed rates due to which action of lateral deformation at the sidewall diminished quickly resulting in lower sized burrs. When the cutter edge reaches point B, the cutter faced maximum uncut material layer. During cutting tool's travel from point B to point C, the uncut layer diminished from maximum to zero. During the rotation of the cutting tool from A to B and then to C, different sized burrs were also formed on the top edge of the transient machined surface. The dimension of the burr size formed around point B may show size effect depending upon the feed per tooth and the cutter edge radius. In this area, around point B, for the uncut layer thickness (feed per tooth) less than the cutter edge radius, burr size increased with the decrease of feed per tooth and for the uncut layer thickness more than the cutter edge radius, burr size decreased with the decrease of feed per tooth. In the down-milling side (B to C), for higher feed rates, cutting action continued to occur till the tool edge is very near to the end of the cut (point C), whereas, for low feed rates, ploughing started to play much earlier before it reached point C. This causes higher sized burrs at the sidewall for lower feed rates at the exit end. The study also revealed that an increase in cutting speed resulted in lower sized burrs. The ease of chip removal with increased temperature in the cutting zone due to an increase in cutting speed might be attributed to this effect. Results obtained in this study showed the contribution of the axial depth of cut was significant in the formation of burrs whence higher axial depths of cut resulted in higher sized burrs. Figure 16 depicts some scanning electron microscope (SEM) images of the top burrs obtained in this study.

Analysis of Cutting Forces in Micro-Cutting of BMG

Cutting forces are important performance measures in micro cutting processes. Cutting force model with sharp edged cutting tool cannot be followed in micro machining as the uncut material thickness is comparable to the cutting edge radius. The influence of the rounded edge needs to be considered in micro cutting to analyse the cutting mechanics. Suitable micro cutting force model captures the cutting phenomenon and accurately predicts the cutting forces. Cutting forces are related to the material removal mechanism. Material deformation in chip formation (e.g., segmentation, etc.) is captured in the variation of the cutting force signatures. Turning of BMG may show the repeated variation of forces though the ideal force signature should be a steady straight line with minor undulations due to variation in the system dynamics. A typical cutting force signature in turning is shown schematically in Figure 17. For a material such as bulk metallic glass which exhibits segmentation in chip formation, each cycle of the repeated variation starts with the initiation of segment generation. Within a cycle, there are marked ups and downs of force profile where the ups indicate the loading and elastic deformation and the downs represent the yielding and shearing off of the secondary shear bands. A number of ups and downs remain in a segment of cyclic variation with gradually increasing peak values and when the segment separates, there is a marked fall in the force profile from the maximum peak which is depicted in Figure 17. This fall represents the shearing of the segment along PSZ. Defects in material micro structure also cause cutting force fluctuations. Improper cutting conditions may lead to excessive cutting force generation, excessive tool deflection, etc. Tool wear causes blunting of the cutting edge resulting in higher cutting forces. Cutting force signatures in micro cutting contain rapid fluctuations and shows erratic behaviour when ploughing is significant and chips are not clearly removed. Various undesirable aspects, such as vibration, tool runout, and tool wear influences process dynamics and are responsible for the erratic nature of the cutting force profile. The quality of the micro machined surfaces is significantly influenced by the abrupt variations in the cutting forces.

Cutting forces in turning of Zr-based and Pd-based BMG were investigated in a study (Fujita et al., 2005) using different cutting tools such as diamond, CBN, ceramics and cermet having 0.4 mm tool tip nose radius, in the cutting speed range of 0.08-2.5 m/s, feed per revolution of 50 μm and radial depth of 50 μm. This study revealed that cutting forces decreased with an increase in the cutting speed for all the cutting tool materials. Thermal softening of the work material was attributed to this effect. Furthermore, the cutting forces were largest for both BMGs using diamond tools. Since diamond has high thermal conductivity compared to other tool materials, it was thought that softening due to the heat generated by cutting is least for the diamond tool tip and this resulted in higher cutting forces. The study also showed that cutting forces in Pd-based BMG were slightly smaller than Zr-based BMG, particularly at lower cutting speeds. Pd-based BMG might have lost strength at lower temperatures compared with the Zr-based BMG during cutting as the glass transition temperature of Pd-based BMG is 100K lower than Zr-based BMG. This study also included other work materials such as steel (JIS SGD-400D) and brass (JIS C3604) for comparative analysis of the cutting forces with BMGs. A notable aspect that was revealed in the analysis is that the cutting forces of the BMGs were much lower than steel at lower cutting speeds though the tensile strength of BMGs is approximately twice that of steel. As BMGs do not generally cause work hardening, BMGs might have lost strength at lower temperatures and exhibited shear flow with smaller shear stresses.

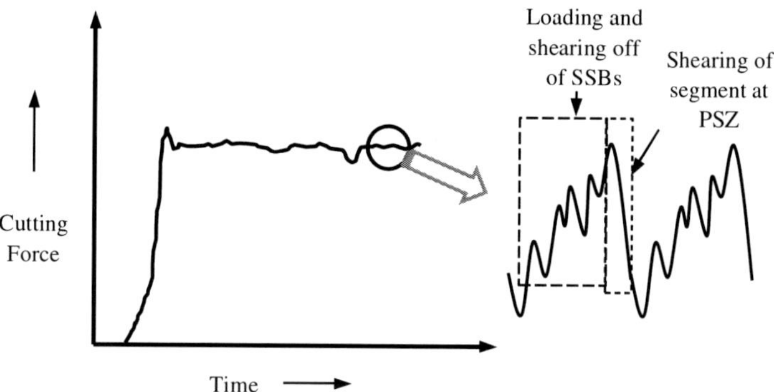

Figure 17. Cutting force signature in turning and force variation related to chip segmentation.

Effects of cutting speed and feed rate were investigated in a study of precision turning of Zr-based BMG (Chen et al. 2017) using the diamond tool (edge radius of 0.33 μm) and PCBN inserts (edge radius of 2.17 μm). Experiments were conducted at cutting speed in the range of 1.05-1.66 m/s, feed rate in the range of 0.5-10 μm per revolution and radial depth in the range of 0.6-20 μm. Cutting forces in this study were found to decrease with an increase in feed rate in the range 1-5 μm at cutting speed of 1.26 m/s and radial depth of 10 μm. However, with further increase in the feed rate in the range of 5-10 μm, cutting forces were found to increase marginally. Cutting forces were found to increase sharply for both types of cutting tools as feed rate decreased below 1 μm. It was notable in the study that the PCBN tool always yielded higher cutting forces than the diamond tool. Analysing cutting forces against cutting speed at a feed rate of 5 μm and radial depth of 10 μm showed that cutting forces were decreased with the increase in the cutting speed. Analysis of specific cutting forces also indicated size effect at lower uncut chip thicknesses. As previously mentioned, strengthening of material at reduced material cut size, ploughing and lack of shear band development and propagation might have been the reason for the size effect. The variation of the cutting forces with feed rate using diamond and PCBN inserts are shown in Figure 18.

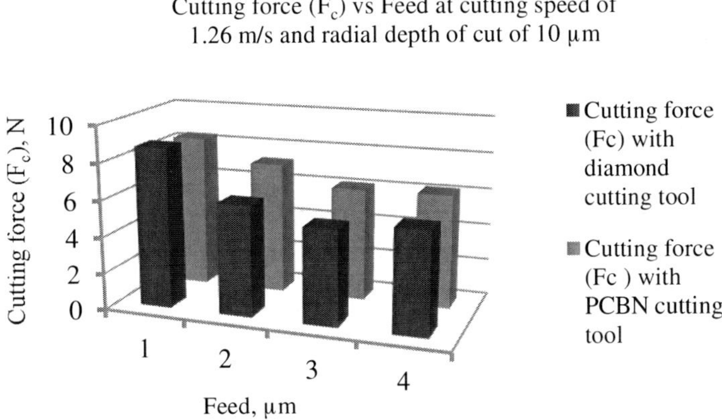

Figure 18. Variation of cutting forces vs feed rate in turning BMG using Diamond and PCBN cutting inserts at cutting speed of 1.26 m/s and radial depth of cut of 10 μm.

In the study on turning of BMG (Bakkal et al., 2004) using tungsten carbide inserts in the cutting speed range of 0.38-1.52 m/s, feed per revolution of 50 μm and radial depth of 0.5 mm, the cutting forces and specific cutting forces were found to decrease with the increase in the cutting speeds. This might be due to the thermal softening of the low thermal conductivity work material at high cutting speeds. Cutting forces were also recorded for other cutting tool materials such as poly-crystalline diamond (PCD) and polycrystalline cubic boron nitride (PCBN). A comparison of the cutting forces from PCD and TiAlN coated PCBN tools of identical geometry showed lower cutting forces with PCBN tools. BMG machined by the WC-PVD tools (rake angle of 20°) had no light emission from the cutting zone and showed much lower forces than the WC-CVD tools (rake angle of 12°).

Axial forces in micro-drilling Zr-based BMG were studied (Zhu et al., 2013) in an experimental investigation using uncoated tungsten carbide micro drills of 0.5 mm diameter in the cutting speed range of 0.21-0.74 m/s and feed rate of 0.31-6.68 μm. The study found an increase in axial forces with an increase in chip load and a marginal decrease of axial forces with an increase in cutting speed. This effect might be explained with the possibility of the BMG sample getting softened as the amorphous structure entered the supercooled liquid state with the rise in temperature at higher cutting speeds to exhibit viscous characteristics when it exceeded the glass transition temperature.

Study of cutting forces in meso-milling Zr-based BMG, stainless steel (SS304) and aluminium alloy (Al6061) using 1 mm diameter uncoated tungsten carbide end mills at a cutting speed of 0.26 m/s and in the feed rate range of 0.25-2 μm showed that cutting forces in Zr-based BMG were higher than Al6061. However, the cutting forces in SS304 were almost twice as high as Zr-based BMG (Bakkal and Naks[idot]ler, 2009). These differences could be attributed to the different work hardening components and deforming mechanisms of the work materials. Micro milling is an important process to create intricate features in micro parts made of bulk metallic glass material. Modeling of cutting forces helps to predict cutting forces at different cutting conditions. Cutting force coefficients are used in the micro cutting model to capture the shearing, ploughing and rubbing effects in cutting. Cutting force coefficients are estimated from some experimental runs valid for a tool-work pair used under specified cutting conditions. Modelling of cutting forces in micro milling BMG (Ray et al., 2020c) revealed that cutting force coefficients bulk metallic glass are comparably higher than that estimated for the materials such as structural steels (Srinivasa and Shunmugam, 2013) and aluminum alloys (Malekian et al., 2009a; Moges et al., 2017). Higher cutting force coefficients indicated higher cutting forces associated with the amorphous metallic alloys than the aforementioned crystalline materials. Increased strengthening at the micro-scale and higher resistance to the deformation for BMG compared to the other materials might be attributed to higher values of the coefficients.

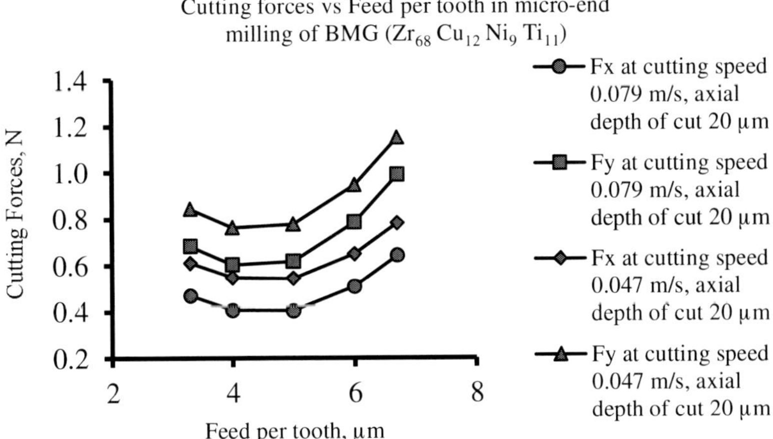

Figure 19.Cutting forces at different cutting conditions; F_x - Cutting force in the direction of feed motion; F_y - Cutting force in the direction perpendicular to the direction of feed motion.

An experimental study of micro milling of BMG sample ($Zr_{68}Cu_{12}Ni_9Ti_{11}$) revealed that in-plane cutting force components (F_x and F_y) increased with decrease in the feed rate at low values of feed rate. On the contrary, at high values of feed rate, these aforesaid force components were found to be increased with increase in feed rate (Ray et al., 2019c). This might have happened due to the intense ploughing action occurring at the cutting edge through the uncut material thickness encountered by the cutting tool is low. At higher feed rates when the feed per tooth was higher than the cutting edge radius, cutting force components increased with the increase in feed rate. The study further revealed that cutting force components decreased with the increase in spindle speed. Increase in cutting speed that increased the temperature in the cutting zone and also increased the strain rate in the localized shear bands of the bulk metallic glass might have an interactive influence on the flow stress and cutting forces. However, in the cutting speed range of this study (0.036-0.09 m/s), the decrease of the cutting forces with the increase in the cutting speed particularly in the range of 0.036-0.063 m/s indicated a predominance of the thermal softening effect. Axial depth of cut significantly influenced cutting force components. Cutting force components increased with the increase in the axial depth of the cut due to the increased engagement of the work material with the cutting tool and the increase in the chip load. The variation of the cutting forces for different cutting conditions is depicted in Figure 19. F_x denotes average peak cutting forces in the direction of feed motion and F_y denotes average peak cutting forces in the direction perpendicular to the feed motion. A notable aspect in micro cutting is that at low chip loads though the cutting forces are reduced, however, the specific cutting forces get significantly increased with the decrease in chip loads. Ploughing and material strengthening effects are proposed to be the main reason for this phenomenon. Material strengthening mechanisms may however differ in crystalline to amorphous materials. In crystalline materials, dislocation mediated slip is scarcely available in the micro level deforming volume. In amorphous materials, the sites of origin for the shear band propagation are greatly reduced in micro-scale volume. A study related to the analysis of specific cutting energy in micro milling of bulk metallic glass (Ray et al., 2020a) indicated that the ratio of uncut material thickness to the cutting edge radius influenced specific cutting forces.

Analysis of Wear of the Cutting Tool in Micro-Cutting of BMG

Tool wear and tool failure are of major concern in micro cutting. Micro tool due to its size, fragility, reduced stiffness and strength is prone to failure under excessive load. Excessive load comes from improper selection of cutting conditions, machining vibrations and cutting tool rotation errors. To avoid tool failure, there is a need for proper dynamic performance of the machine tool, use of cutting tools with accurate geometry, carrying out operations with appropriate cutting conditions and use of work material with homogeneity. Tool wear influences surface finish, burr formation and part accuracy. Tool wear of the end mill in micro end milling lowers the actual radial and axial depths of cut in the material affecting profile shape and accuracy. It also lowers the depth of groove on the surfaces generated due to feed motion, causing excessive smearing and ploughing of material and depositing material over the feed marks. In commercially available micro milling tools, small size and fragility make the micro tool susceptible to failure. Tool wear and failure in micro cutting is not uniform and unpredictable to some extent. Tool wear monitoring system may be integrated with the machining system to acquire extensive information on the machining process; system dynamics and tool wear related phenomena. Researchers have used various methods for tool wear monitoring and analysis. Some of the methods are processing of acoustic emission signals (Jemielniak and Arrazola, 2008), analysis of scanning electron microscopy images (Tansel et al., 1998), neuro fuzzy method using signals from sensors such as accelerometer, force sensor and acoustic emission sensor (Malekian et al., 2009b), processing of wear images (Zhu and Yu, 2017), etc. Limited studies were reported regarding cutting tool wear related to the machining of BMG. A study on micro turning Zr-based BMG with single crystal diamond tool and PCBN inserts revealed that adhesive wear was significantly present in the cutting process (Chen et al., 2017). X-ray photoelectron spectroscopy (XPS) studies on the cutting tools revealed that XPS spectra contained peaks corresponding to Zr-O as a result of oxidation of the adhered BMG material due to higher cutting temperatures. However, the XPS spectra corresponding to the diamond cutting tool had additional peaks of Zr-C bonds from which it can be concluded that carbon atoms of the diamond tool reacted with the zirconium of the BMG material thereby forming ZrC. For the PCBN tool, the cubic boron nitride grains were rapidly depleted by adhesive wear as the adhered work material split away off the cutting tool edge. In contrast to this, diamond tool had less adhesive wear resulting in better surface quality as compared to the PCBN tool. Tool wear in the diamond tool was mostly due to a slow diffusive mechanism that resulted in less wear than the PCBN tool.

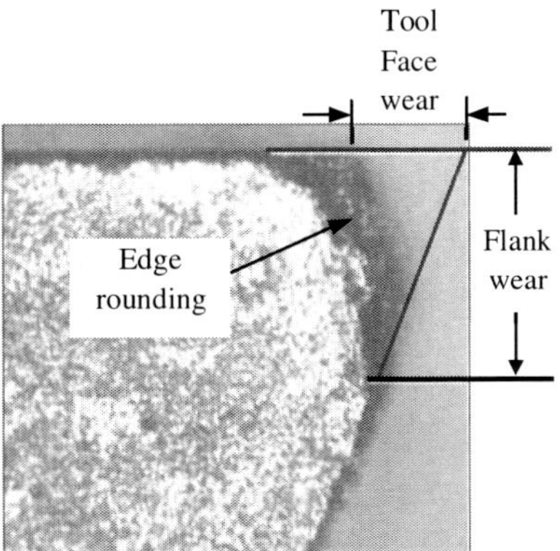

Figure 20. Different dimensions of cutting tool edge wear.

A study on micro-drilling of Zr-based BMG revealed that considerable tool wear occurred both on the face and flank side (Zhu et al., 2013). Tungsten carbide micro drill of diameter 0.5 mm was used in the study to estimate tool wear against a number of micro holes drilled at a cutting speed of 0.33 m/s and feed per flute of 1.25 μm. A cumulative flank wear (width of flank wear land) of 20 μm and rake wear (width of cratering) of 30 μm were observed against the total number of 100 micro holes drilled. A notable aspect observed in this study was that in the case of flank wear, a distinct 'wear in' period was followed by gradual wear, whereas in the tool rake face side, there was a hyperbolic steady rise in tool wear. Moreover, the wear at the periphery was higher than that at the centre. Wear on the tool face was attributed mainly due to the diffusion of cobalt into the work material. On the other hand, flank wear was mainly attributed due to abrasion of the cutting edge. Turning experiments on Zr-based BMG at cutting speeds in the range of 0.38-1.52 m/s and at a feed rate of 50 μm (higher than normally considered in micro cutting) with different tools namely WC (Tungsten carbide), PCBN (Polycrystalline cubic boron nitride) and PCD (Polycrystalline diamond) (Bakkal et al., 2004) revealed that melted chip segments were adhered to the wear out cutting edges of WC tools. This indicates that WC tools sustained considerable adhesive wear. This study also indicated that PCBN and PCD tools had chipping of cutting edges in machining BMG workpieces. All the above mentioned studies indicated that the cutting edges had edge rounding and flank wear development as shown in Figure 20.

Conclusion

With the advancement of science and technology and emerging requirements for smaller, reliable, efficient and precise components needed for micro devices, there has been an increasing demand to manufacture miniaturized products with micro features. The use of Bulk Metallic Glass in recent days is finding applications in several areas ranging from sports goods to biomedical devices. Many of these applications include micro-parts made of bulk metallic glasses. Micro-cutting is one of the most reliable processes to create micro-features. For effective utilization of micro cutting bulk metallic glass material in industrial practices, a thorough understanding of the micro-cutting process mechanism, critical aspects of chip formation and the influence of the cutting condition on the performance measures need to be addressed. This chapter discussed several aspects in depth such as fundamental aspects of the micro cutting process, analysis of chip formation, analysis of process performance measures that include analysis of the quality characteristics of micro machined features, analysis of cutting forces, and analysis of the wear of cutting tool related to micro-cutting BMG material. The discussion presented here will help the researchers and scientists to understand the micro-cutting process mechanisms better to utilize precision and micro-cutting processes in shaping micro components made of bulk metallic glass.

References

Abdelmoneim MEs, Scrutton RF. Tool edge roundness and stable build-up formation in finish machining. *Journal of Engineering for Industry* (1974) 96(4):1258–1267.

Ahmed W, Jackson MJ. Emerging nanotechnologies for manufacturing. Oxford, UK: *Elsevier;* 2014.

Albrecht P. New developments in the theory of the metal-cutting process: part i. the ploughing process in metal cutting. *Journal of Engineering for Industry* (1960) 82(4):348–357.

Bakkal M, Nakş[idot]ler V. Cutting mechanics of bulk metallic glass materials on meso-end milling. *Materials and Manufacturing Processes* (2009) 24(12):1249–1255.

Bakkal M, Shih AJ, Scattergood RO. Chip formation, cutting forces, and tool wear in turning of Zr-based bulk metallic glass. *International Journal of Machine Tools & Manufacture* (2004) 44(9):915–925.

Bakkal M, Shih AJ, McSpadden SB, Liu CT, Scattergood RO. Light emission, chip morphology, and burr formation in drilling the bulk metallic glass. *International Journal of Machine Tools & Manufacture* (2005) 45(7-8):741–752.

Balasubramaniam R, Suri VK. 'Diamond Turn Machining' in VK. Jain (ed.) *Introduction to Micromachining*, New Delhi: Narosa Publishing House; 2011.

Bang Y, Herschorn S, Oh S. 5-axis micro milling machine for machining micro parts. *Int. J. Adv. Manuf. Technol.* (2004) 25(9-10):888–894.

Basu J, Ranganathan S. Bulk metallic glasses: A new class of engineering materials. *Sadhana* (2003) 28(3-4):783–798.

Chae J, Park SS, Freiheit T. Investigation of micro-cutting operations. *International Journal of Machine Tools and Manufacture* (2006) 46(3-4):313–332.

Chen M. A brief overview of bulk metallic glasses. *NPG Asia Materials* (2011) 3(9):82–90.

Chen XD, Xu J, Zhu Y, Tian R, Shu X. Micro-machinability of bulk metallic glass in ultra-precision cutting. *Materials and Design* (2017) 136:1–12.

Connolly R, Rubenstein C. The mechanics of continuous chip formation in orthogonal cutting. *International Journal of Machine Tool Design and Research* (1968) 8(3):159–187.

Design guide 5.0. Available at: https://liquidmetal.com/wp-content/uploads/2021/05/LQM_DesignGuide.

Dhale K, Banerjee N, Singh RK, Outeiro JC. Investigation on chip formation and surface morphology in orthogonal machining of Zr-based bulk metallic glass. *Manufacturing Letters* (2019) 19:25–28.

Ding F, Wang C, Zhang T, Zheng L, Zhu X, Li W, Li L. Investigation on chip deformation behaviors of Zr-based bulk metallic glass during machining. *Journal of Materials Processing Technology* (2020) 276:116404–116404.

Dornfeld D, Min S, Takeuchi Y. Recent Advances in Mechanical Micromachining. *CIRP Annals*, (2006) 55(2):745–768.

Duan G, Wiest A, Lind ML, Li J, Rhim WK, Johnson WL. Bulk Metallic Glass with Benchmark Thermoplastic Processability. *Advanced Materials* (2007) 19(23):4272–4275.

Filiz S, Conley CM, Wasserman MB, Ozdoganlar OB. An experimental investigation of micro-machinability of copper 101 using tungsten carbide micro-endmills. *International Journal of Machine Tools and Manufacture* (2007) 47(7-8):1088–1100.

Fujita K, Morishita Y, Nishiyama N, Kimura H, Inoue A. Cutting Characteristics of Bulk Metallic Glass. *Materials Transactions* (2005) 46(12):2856–2863.

Han DH, Wang G, Li JC, Chan KM, To S, Wu FF, Gao Y, Zhai Q. Cutting Characteristics of Zr-Based Bulk Metallic Glass. *Journal of Materials Science & Technology* (2015) 31(2):153–158.

Huo D, Cheng K. Overview of micro-cutting, in Cheng k, Huo D. (Eds), Microcutting: Fundamentals and Applications, Chichester, UK: *John Wiley & Sons Ltd.*; 2013.

Ikawa N, Shimada S, Tanaka H. Minimum thickness of cut in micromachining. *Nanotechnology* (1999) 3(1):6–9.

Inamura T, Takezawa N, Kumaki Y. Mechanics and Energy Dissipation in Nanoscale Cutting. *Annals of the CIRP* (1993) 42(1):79–82.

Jang D, Greer JR. Transition from a strong-yet-brittle to a stronger-and-ductile state by size reduction of metallic glasses. *Nature Materials* (2010) 9(3):215–219.

Jemielniak K, Arrazola PJ. Application of AE and cutting force signals in tool condition monitoring in micro-milling. *CIRP Journal of Manufacturing Science and Technology* (2008) 1(2):97–102.

Jiang MQ, Dai LH. Formation mechanism of lamellar chips during machining of bulk metallic glass. *Acta Materialia* (2009) 57(9):2730–2738.

Kim CJ, Bono M, Ni J. Experimental analysis of chip formation in micro-milling. In: *NAMRC XXX, West Lafayette, Indiana. Dearborn, MI 48121: Society of Manufacturing Engineers* (2002) (1):1–8.

Kim CJ, Mayor JR, Ni, J. A Static Model of Chip Formation in Microscale Milling. *J. Manuf. Sci. Eng.* (2004) 126(4):710–718.

Kim JH, Kim D. Theoretical analysis of micro-cutting characteristics in ultra-precision machining. *J. Mater. Process. Technol* (1995) 49(3-4):387–398.

Kumar G, Tang HX, Schroers J. Nanomoulding with amorphous metals. *Nature* (2009) 457(7231):868–872.

Kuzmin OV, Pei Y, Chen C, Hosson DJ. Intrinsic and extrinsic size effects in the deformation of metallic glass nanopillars. *Acta Materialia* (2012) 60(3):889–898.

Lekkala R, Bajpai V, Singh RK, Joshi SS. Characterization and modeling of burr formation in micro-end milling. *Precision Engineering* (2011) 35(4):625–637.

Liu XJ, DeVor RE, Kapoor SS, Ehmann KF. The Mechanics of Machining at the Microscale: Assessment of the Current State of the Science. *Journal of Manufacturing Science and Engineering-transactions of The ASME* (2004) 126(4):666–678.

Liu K, Melkote SN. Material Strengthening Mechanisms and Their Contribution to Size Effect in Micro-Cutting. *Journal of Manufacturing Science and Engineering* (2005) 128(3):730–738.

Lucca DA, Seo YW, Komanduri R. Effect of Tool Edge Geometry on Energy Dissipation in Ultraprecision Machining Annals of the CIRP (1993) 42(1):83-86.

Malekian M, Park SS, Jun MBG. Modeling of dynamic micro-milling cutting forces. *International Journal of Machine Tools & Manufacture* (2009a) 49(7-8):586-598.

Malekian M, Park SS, Jun MBG. Tool wear monitoring of micro-milling operations. *Journal of Materials Processing Technology*, (2009b) 209(10):4903–4914.

Maroju NK, Jin X. Mechanism of Chip Segmentation in Orthogonal Cutting of Zr-Based Bulk Metallic Glass. *Journal of Manufacturing Science and Engineering* (2019) 141(8):081003/1-13.

Moges TM, Desai KA, Rao PVM. On modeling of cutting forces in micro-end milling operation. *Machining Science and Technology*, (2017) 21(4):562–581.

Pan CT, Wu T, Chen M, Chang YH, Lee, CS, Huang JC. Hot embossing of micro-lens array on bulk metallic glass. *Sensors and Actuators A* (2008) 141(2):422–431.

Ray D, Puri AB. Mechanical Micro-milling: A Broader perspective of the process, in Dikshit MK, Pathak VK, Puri AB, Davim JP (eds.), Futuristic Manufacturing Perpetual Advancement and Research Challenges. London: CRC Press; 2023.

Ray D, Puri AB, Hanumaiah N, Halder S. Modeling of top burr formation in micro-end milling of Zr-based bulk metallic glass. *Journal of Micro and Nano-Manufacturing* (2019a) 7(4):041004.

Ray D, Puri AB, Hanumaiah N, Halder S. An experimental analysis on the formation of top burrs in micro-end milling. *International Journal of Mechanical and Production Engineering* (2019b) 7(7):17-23.

Ray D, Puri AB, Nagahanumaiah (2019c). Investigation on cutting forces and surface finish in mechanical micro milling of Zr-based bulk metallic glass. *Journal of Advanced Manufacturing Systems* (2019c) 18(1):113–132.

Ray D, Puri AB, Hanumaiah N, Halder S. Analysis on specific cutting energy in micro milling of bulk metallic glass. *The International Journal of Advanced Manufacturing Technology* (2020a) 108(1-2):245–261.

Ray D, Puri AB, Hanumaiah N. Experimental analysis on the quality aspects of micro-channels in mechanical micro milling of Zr-based bulk metallic glass. *Measurement*, (2020b) 158:107622.

Ray D, Puri, AB, Hanumaiah N, Halder S. Mechanistic modelling of dynamic cutting forces in micro end milling of Zr-based bulk metallic glass. *International Journal of Machining and Machinability of Materials* (2020c) 22(6):527–550.

Sarwar M, Thompson PJ. Cutting action of blunt tools in Davies BJ. (Eds) Proceedings of the Twenty-second International Machine Tool Design and Research Conference. London: Red Globe Press; 1982.

Schroers J. Processing of Bulk Metallic Glass. *Advanced Materials* (2009) 22(14):1566–1597.

Schroers J, Johnson WL Ductile Bulk Metallic Glass. *Physical Review Letters* (2004) 93(25):889-898.

Srinivasa YV, Shunmugam MS. Mechanistic model for prediction of cutting forces in micro end-milling and experimental comparison. *International Journal of Machine Tools and Manufacture*, (2013) 67:18–27.

Stephenson DA, Agapiou JS. *Metal Cutting Theory and Practice*. Boca Raton, FL: CRC Press; 2016.

Tansel IN, Rodriguez O, Trujillo M, Paz E, Li, W. Micro-end-milling-I. Wear and breakage. *International Journal of Machine Tools and Manufacture* (1998) 38(12): 1419–1436.

Ueda K, Manabe K. Chip Formation Mechanism in Microcutting of an Amorphous Metal. *Annals of the CIRP* (1992) 41(1):129–132.

Waldorf DJ, DeVor RE, Kapoor SG. An Evaluation of Ploughing Models for Orthogonal Machining. *Journal of Manufacturing Science and Engineering* (1999) 121(4):550–558.

Wang WH, Dong C, Shek CH. Bulk metallic glasses. *Materials Science and Engineering: R: Reports* (2004) 44(2-3):45–89.

Weck M, Hennig JC, Hilbing R. Precision cutting processes for manufacturing of optical components. *Proc. SPIE 4440, Lithographic and Micromachining Techniques for Optical Component Fabrication* (2001) 4410:145-151.

Wert JA, Thomsen C, Jensen RD, Arentoft M. Forming of bulk metallic glass microcomponents. *Journal of Materials Processing Technology (2009)* 209(3):1570–1579.

Wu X, Li L, He N, Yao C, Zhao M. Influence of the cutting edge radius and the material grain size on the cutting force in micro cutting. *Precision Engineering* (2016) 45:359–364.

Zhu J, Kim HJ, Kapoor SG. Microscale Drilling of Bulk Metallic Glass. *Journal of Micro and Nano-Manufacturing* (2013) 1:041004.

Zhu K, Yu X. The monitoring of micro milling tool wear conditions by wear area estimation. *Mechanical Systems and Signal Processing* (2017) 93:80–91.

Chapter 3

Progress on Functional Properties of [Ni-(Mo/Cr)-Si]:[Nb/Ti] Bulk Metallic Glass Systems

Gayatri Tanuja Guddla[1]
Vamsi Krishna Katta[2]
Bharadwaj Somayajula[2]
and Balaji Rao Ravuri[3,*]

[1]Department of Mechanical Engineering, New Horizon College of Engineering, Bangalore, India
[2]Department of Physics, School of Science, GITAM (Deemed to be University), Hyderabad, India
[3]Department of NanoSciences and Materials, Central University of Jammu, Jammu, India

Abstract

Bulk Metallic Glasses (BMGs) are the newest functional glass metallic systems, having superior mechanical and thermal properties when compared to their crystalline counterparts with low critical cooling rates. The current chapter highlights the significance of bulk metallic glasses (BMG) based on [Ni-(Mo/Cr)-Si]:[Nb/Ti] systems and their distinct properties such as avoiding sulphuric acid embrittlement, overcoming strain cracking issues, preventing pitting, and enhancing corrosive resistance, all of which are obligated for aerospace and lightweight vehicle implementations. Advances in [Ni-(Mo/Cr)-Si]:[Nb/Ti] are still ongoing to improve towards industrial application with research being

* Corresponding Author's Email: balaji.nsm@cujammu.ac.in.

In: Properties and Uses of Metallic Glass
Editor: Shiv Prakash Singh
ISBN: 979-8-89113-569-7

carried out on the aspects of corrosion properties. Thermal, mechanical, and potentiometric parameters variation in samples can be explained using microstructure changes.

Keywords: bulk metallic glasses, glass forming abilities, thermal properties, mechanical properties

Introduction

Bulk metallic glasses also known as BMGs are a special class of functional materials with low critical cooling rates, mechanical and thermal capabilities in comparison to their crystalline counterparts. In fact, before the first metallic glass system, researchers were dubious about whether they could make metallic glasses in bulk since they demand substantially lower cooling rates, than AuSi(M. Chen, 2011; Klement, Willens, & Duwez, 1960). Over decades, researchers from all fields including engineering, medical, aerospace-space applications, automobiles, precise gear, cutting-edge robotics, optical mirror devices, surface coatings, and light-weight vehicle applications(Inoue, 2000; Schuh, Hufnagel, & Ramamurty, 2007; Yavari, Lewandowski, & Eckert, 2007).

An optimal BMGs material would possess properties with the benefits of ceramics and metals such as:

- Having toughness more than ceramic materials.
- Having wear resistance higher than any metal.
- Having low processing temperatures suitable for forming required shapes.
- Easily Machinable.
- Robust to any extreme environments.
- Thermal stability,
- Hardness,
- Great corrosion resistance,
- High glass-forming
- Cast-ability abilities
- Low critical cooling rates
- Low crystalline flaws

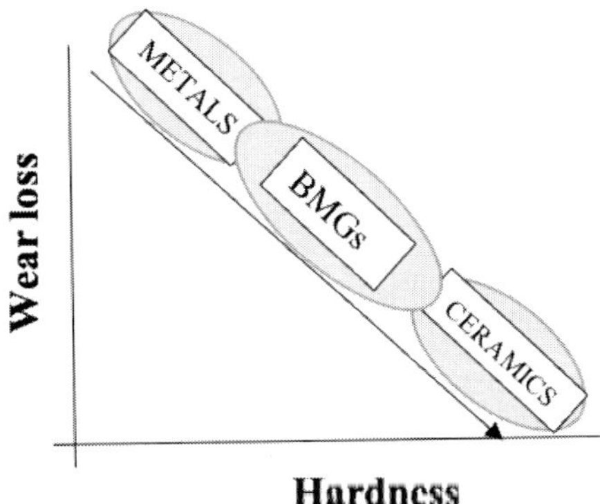

Figure 1. Schematic showing the importance of BMGs.

Crystalline materials exhibit very poor wear whereas ceramic materials are too brittle. BMGs are found to be the best solution to above mentioned requirements that can showcase excellent strength and fracture toughness which are not possible with regular metals, that attracts use in many applications. They offer two times more toughness than ceramic materials. Also, earlier seen shortcomings of materials such as apparent brittleness, and poor fatigue resistance are not at all a big problem after the invention of BMGs. This can be clearly explained with the schematic diagram of BMG, shown in Figure 1. From the literature survey, we have already seen the methods involved in the synthesis, advantages, disadvantages, and applications of BMGs. Also, we have seen numerous varieties of combinations and compositions of various metals involved in making BMGs.

A decade's worth of research has been conducted by numerous groups on the structural and functional characteristics of a variety of elements such as magnesium, zirconium, titanium, iron, copper, and nickel BMG composites (Brink, Adjaoud, & Albe, 2016; Cheng, Sheng, & Ma, 2008; X. Gu, Poon, & Shiflet, 2007). It was often recommended to use multi-element components with varied compositions possessing atomic size mismatches (Hofmann, 2013). Amorphous metallic materials have different physical, mechanical, and chemical properties from other conventional metallic materials due to the absence of long-range order. A large range of multi-component alloys was developed by conventional casting techniques, which required significantly lower critical cooling rates with metal base systems such as Ti-, Ni-, Zr-, Fe-, Co-, etc., (Inoue & Kong, 2016; Kawamura, Nakamura, & Inoue, 1998). The first BMGs were created about thirty to forty years later, based on Pd. The structural, mechanical, and corrosion properties of BMGs have been studied and reported for a wide range of BMGs; however, research is still being conducted to enhance these properties to meet the needs of contemporary industrial applications (Inoue, 1999; Khalifa, 2009; Z. Lu & Liu, 2004; E. Park, Kim, Ohkubo, & Hono, 2005; Peker & Johnson, 1993; Sarac, 2015; C Suryanarayana & Inoue, 2011). BMGs are often expected to exhibit inhomogeneous deformation as a result of localized shear bands that soften under strain and temperature, resulting in controlled ductility (Nieh & Wadsworth, 2006; C Suryanarayana & Inoue, 2013). The majority of BMGs made up of four or more elements, exhibit a wide super-cooled zone with atomic size mismatches greater than 13% and possess negative heat values during mixing (Inoue, Shen, & Nishiyama, 2008; Cury Suryanarayana & Inoue, 2017).

The BMG composite materials can be produced by melting and solidification processes, and display distinct physical, chemical, mechanical, and thermal strengths due to medium-range organized domains. In order to meet the expectations of the industry, alternative systems and methods must be developed (Cao, Wu, Wang, Liu, & Lu, 2016; Tan, Jiang, Zhang, Wu, & Lin, 2008; Yang et al., 2019). Together, these findings demonstrated that certain characteristics, such as thermal stability, yield, and compressive strengths (σ_y, σ_f), for instance, could not be achieved at a satisfactory level. Further research established that the only realistic way to increase the glass-forming ability of BMGs is by minor alloying of the base glass system with metals (N. Chen, Martin, Luzguine Luzgin, & Inoue, 2010; Guojian Hao, Ren, Zhang, & Lin, 2009).

Ni-based BGMs (Hitit, Şahin, Öztürk, & Aşgın, 2015) suggested the ability of nickel metal to be electroplated with other metals and further used as protective coatings. By adding molybdenum (Mo) metal to nickel alloys, corrosion resistance, and alloy strength can be significantly improved (G. Cao et al., 2019; Filipecka, Pawlik, & Filipecki, 2017; He, Shang, Ma, & Xu, 2012; Liang, Shen, & Sun, 2006; Shao et al., 2018; W. H. Wang, Dong, & Shek, 2004; X. Zhang, Ma, & Xu, 2006). The presence of Silicon (Si) with molybdenum (Mo) with nickel (Ni) improves anticorrosive properties even in sulphuric conditions. In comparison to other transition metal elements, titanium (Ti) exhibits utilitarian features such as increased thermal, mechanical, and corrosive resistance. Silicon (Si), a key constituent ingredient, is essential to the creation of BMGs glass. Liang WZ et al. (2016) research suggested that silicon possesses higher GFA with strain and plastic deformation. Niobium (Nb) doping in BMGs has several benefits, including (i) increasing GFA and (ii) strengthening the BMGs' mechanical, thermal, and physicochemical properties. However, it is widely known that even a small amount of alloying can have a big impact on how well BMGs work (C. L. Chen & Huang, 2014; Gandi, Chinta, Ghoshal, & Ravuri, 2019; Lin & Spaepen, 1982; S. Liu et al., 2016).

Our research in BMGs matrices with the formula $[Ni\text{-}(Mo/Cr)\text{-}Si]_{100-x}:[Ti/Nb]_x$ (x = 0 - 15 at.%) using a high energy ball milling as a synthesis technique. Considering their outstanding properties for structural, mechanical, thermal, and corrosion, one best sample from each series is chosen and assigned the designations $NMST_6$, $NMSNb_{10}$, $NCST_8$, and $NCSNb_6$. $NMST_6$ is $(Ni_{75}Mo_{15}Si_{10})_{94}Ti_6$ whereas $NCST_8$ is $(Ni_{75}Cr_{15}Si_{10})_{92}Ti_8$. The findings showed that combining the right amounts of Ti/Nb content into the current BMG network helped to increase corrosion resistance, avoid embrittlement, solve strain cracking issues and reduce pitting. The structural variation and glass-modifying properties brought about by the substitution of Ti/Nb in the proper quantities may be understood in the context of all of the targeted samples' variations in thermal (T, T_{rg}), mechanical (σ_y, σ_f, hardness), and potentiometric (E_c, i_c, E_p, i_p) parameters. Titanium (Ti) has demonstrated excellent resistance to age hardening and cracking brought on by polythionic acid (Inoue, Kong, Zhu, & Al Marzouki, 2017; Katta et al., 2021; Stolle & Ranu, 2014). It is a metal with a low specific gravity and when added it to the (Ni-Mo-Si) BMG metal, the alloy prevents strain cracking, enhances fatigue resistance, and further acts as a deoxidizer. This development is ideal for lightweight alloy applications in jet engines and airplanes, even in high-pressure and temperature environments(Liaw & Miller, 2007; Suñol, Clavaguera, & Clavaguera Mora, 2001). Additionally, this experiment demonstrates that Niobium (Nb), even at a concentration of 0.1%, can greatly improve performance and provide outstanding resistance to oxidation and corrosion. Method of preparation (G. Guddla, Katta, Gandi, Ambadipudi, & Ravuri, 2021; G. T. Guddla, Ambadipudi, Katta, Katari, & Ravuri, 2022; G. T. Guddla, Ambadipudi, Yenduva, Katta, & Ravuri, 2022; G. T. Guddla, Gandi, Ambadipudi, & Ravuri, 2021; Manjunatha et al., 2022) plays a vital part in optimizing the properties and among them, ball milling method is said to be a versatile and simple method. High energy ball milling method was used to synthesize $[Ni\text{-}Mo\text{-}Si]_{100-x}:Nb_x$ BMG glass systems. In the ball milling method, the amount of crystallization caused by stress and milling medium, as well as contaminants in the powder mixture, can be adjusted. In addition to this composition variation and microstructure condition determines plastic deformation, particle welding, and fracture morphology.

Amorphous vs Crystalline Solids

Solids are known to have extremely well-organized structures with millions of their atoms arranged in a predictable way, including crystals of salt and sugar. The molecules will align themselves in a particular configuration and are rigidly bonded in a crystalline solid. In contrast to the strictly bound and disordered glass molecules, the molecules in liquids are not rigorously bound. Amorphous substances, such as glass, rubber, cotton candy, butter, etc., solidify too quickly for an orderly structure to develop. The recognized state of matter for glass is an amorphous solid, which is a supercooled liquid and lies in the transition region between the two. Glass is a solid amorphous body that contains all types of atomic bonds, including ionic, covalent, metallic, and Vander Waals. They are the most challenging to produce due to the non-directional nature of the metallic link and can only be done so by using very fast cooling rates (106 $K.s^{-1}$). As a result, metallic glasses are originally created via melt spinning or splat quenching processes as thin foils or ribbons (50 m).

Ball Milling

Many scientific researchers have been on-going towards developing a synthesis method with less cost and making sub-micron fine powders with controlled distribution of size. Among these, the Ball Milling technique turned out to be the significant one for making sub-micrometre powders. Inoue, et al., found this technique to be cost-effective, simple, and can be used with a wide range of materials. Ball mills are now extensively being used for mechanical alloying process to produce alloys from powders, by grinding and cold welding (Churyumov, Bazlov, Tsarkov, Solonin, & Louzguine Luzgin, 2016; Feng et al., 2019).

The term "Ball Mill" refers to a milling device that is also known as a "pebble mill" or "tumbling mill." It comprises a hollow cylindrical jar with balls inside which is set on a rotating metal frame. Higher velocities are produced as a result of this rotation, which transfers a significant amount of kinetic energy from the balls to the sample and causes it to be reduced to a fine powder. By impact and attrition, this occurs. The milling duration in this operation is a crucial variable that can affect the powder's structure, grain development, morphology, and physical qualities. Additionally, the rotation speed and ball sizes that fill 30 to 40% of the mill's capacity are crucial components. So, a Ball Mill is a major piece of equipment for crushing and producing fine powders from materials and is widely used in many applications including glass ceramics, cement, silicates, fertilizer, refractories, etc. (J. Gu et al., 2019; E. Park, Chang, Kim, Ohkubo, & Hono, 2006; E. Park & Kim, 2005; J. Park et al., 2011; D. Wang et al., 2004) Also, used for ore-dressing of ferrous and non–ferrous materials. Size, hardness, density and composition are the key factors of grinding media (balls). Inert shield gas is also used to fill the grinding chamber in order to prevent any explosive reactions inside the mill and oxidation. The ball mills mostly used in laboratories are Planetary Ball mills that are smaller compared to the common ball mills. It consists of 1–4 grinding jars, called vials that are arranged eccentrically on a common base called the sun wheel. These vials and sun wheel will rotate in opposite directions developing Coriolis forces and create superimposed rotational moments that release high dynamic energies used for rapid fine crushing of all kinds of materials (hard, soft, brittle, fibrous) with end fineness < 1μm. They have exchangeable jars easy to clean n handle. Both dry and wet grinding are possible with this kind of ball mill as shown in Figure 2. (Guo et al., 2012; Hasegawa, Takeuchi, Kato, & Inoue, 2004; L. Li, Liu, Zhao, Cai, & Yang, 2014; C. T. Liu & Lu, 2005; Löffler, 2003).

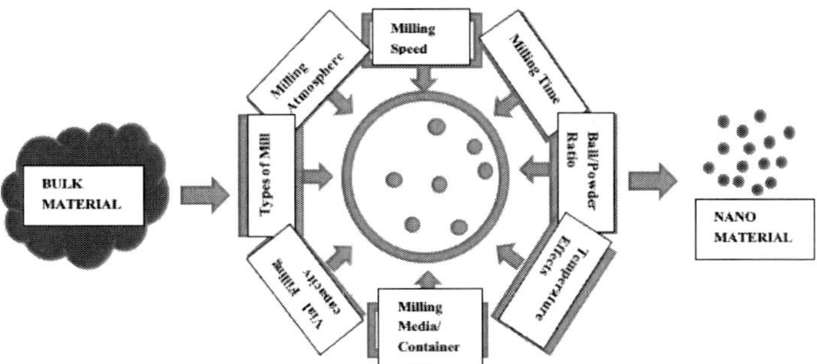

Figure 2. Planetary Ball Milling Equipment.

Factors Influencing the Degree of Milling in Ball Milling Apparatus

- The nature of the balls used for milling.
- Speed of the vials in rpm.
- The size, number, and density of balls used in vials set as per feed material to balls ratio (Usually taken as 1:10 ratio). The size and density of the balls should be substantially more than the material to be ground.

- Time of milling operation.
- Feed rate in the vials.
- Wet-grinding is recommended while grinding flammable materials in order to avoid an explosion powdered form. The wet-grinding media can be hexane solution, toluene, etc.
- In case, less contamination is the main criteria of the final product, then the stainless-steel dust of the balls, if any, can be separated by magnetic means.

Advantages of Ball Milling

- Grinding medium and low cost of installation.
- As the jars are completely and securely enclosed, it is safe for milling of toxic materials.
- We can produce very fine powder of particle size <= 10 microns. The fineness can also be adjusted by adjusting the ball diameter.
- It can be used for prolonged operation of continuous hours. This apparatus is used for both continuous and batch operation.
- By choosing the proper size and density of balls, this equipment can be used for materials of all degree of hardness.
- Used in a wide range of applications including milling ores and grinding of drugs in pharmaceutical fields.

Disadvantages of Ball Milling

- There are chances of contamination of the final product which may occur due to wear and tear of vial material or from balls.
- The entire operation is noisy.
- The milling time is pretty high which requires consistent follow-up and patience.
- After each and every use, it requires much effort and care for cleaning of vials.

Thermal Analysis on BMGs

DTA is a thermal analysis method that calculates the temperature difference (ΔT), or weight loss (%) as a function of temperature between a sample and standard reference material. The reference and sample materials are set up in an identical thermal environment, and as the sample warms up or cools down, the temperature difference between the two is recorded. In DTA, the sample and reference are usually placed in separate furnace chambers, and the temperature is gradually raised at a steady rate while the sample and reference's temperature difference is continually monitored. When a thermal event occurs in the sample, such as a phase transition, melting, or crystallization, it is reflected as a sharp peak in the DTA curve. The area under the peak in the DTA curve is proportional to the heat flow associated with the thermal event. DTA can provide useful information about a material's thermal behaviour, such as melting point, crystallization temperature, phase transitions, and thermal stability. Additionally, it can be used to look into the kinetics of phase changes and how additives or impurities affect a material's thermal behaviour (Bhadeshia, 2002; Busch, 2000; Inoue, Zhang, & Masumoto, 1990).

Working Principle of DTA

When a substance is heated or cooled at a predetermined uniform pace, the temperature of the subject material is measured, recorded, and then compared to the thermally inert material, such as alumina. Compared to the range that can be determined using TG, the temperature range that can be monitored during DTA is much greater. The endothermic or exothermal departure from the baseline during DTA serves as a signal for these alterations. Since DTA is a dynamic method, standardization of every element of the technique is necessary to get repeatable results. These include the specimen's preparation, particle size and packing, dilution, and the type of inert diluent used (D. Cao et al., 2019; Debenedetti & Stillinger, 2001; Louzguine & Inoue, 2002; Pang, Zhang, Asami, & Inoue, 2002). With thermal changes in the sample, changes are often evident on the basis of crystalline structure, decomposition reactions, oxidation, and reduction reactions. Crystallisation, oxidation, and certain decomposition processes are examples of exothermic reactions, whereas phase shifts, dehydration, reduction, and some breakdown processes are examples of endothermic reactions.

Physical Characteristics of DTA for BMGs

The endothermic profiles that are the key indicator of a glass transition area are generally present in all of the BMG samples. This is followed by a super-cooled liquid phase and an exothermic reaction. All thermal parameters, including the reduced glass transition temperature (T_{rg}) for all synthesized

BMG samples, the glass transition temperature (T_g), the onset or crystallization temperature (T_c), the melting point temperature (T_m), the liquids temperature (T_l), and the super-cooled liquid region ($\Delta T = T_c - T_g$). GFA parameter ($\gamma = T_c/(T_g + T_l)$, and reduced glass transition temperature (T_{rg}) for all synthesized BMG samples. After 30 hours of ball milling, Figure 4 displays the DTA profiles of all as-prepared $NMST_6$, $NMSNb_{10}$, $NCST_8$, and $NCSNb_6$ BMG samples, while Table 2 displays the DTA parameters of all as-prepared BMG samples.

Glass Transition Temperature (T_g)

The glass transition temperature (T_g) of bulk metallic glasses (BMGs) refers to the temperature at which the material's amorphous structure begins to transition from a supercooled liquid state to a glassy solid state. BMGs are metallic alloys with a disordered atomic structure, similar to that of a liquid or glass, rather than long-range crystalline order. BMGs have high strength, good corrosion resistance, and other desirable properties due to their unique structure. The glass transition temperature of BMGs is determined by their chemical composition as well as their processing history. Because of their high atomic packing density and complex atomic arrangements, BMGs have lesser T_g values than conventional glasses in general. The glass transition temperature is a critical parameter in the processing and engineering of BMGs because it affects their mechanical properties and ability to be moulded or shaped. Understanding and controlling the T_g of BMGs is thus critical for the development of new applications and materials based on these promising alloys. The $NCSNb_6$ sample gave the lesser T_g value (613 K) and formed good glasses state among all other BMGs as mentioned in Table 1 (Du, Huang, Liu, & Lu, 2007; Inoue, Zhang, & Masumoto, 1993; Men, Hu, & Xu, 2002; Russo, Romano, & Tanaka, 2018).

Crystallization Temperature (T_c)

Bulk metallic glasses' (BMGs') condensation temperature is influenced by the alloy's chemical makeup. BMGs typically resist crystallization better than their crystalline counterparts because the glassy form lacks a long-range ordered structure. The exothermic peak's start temperature on the DTA curve (Figure 3) serves as the standard measurement for the crystallization temperature. Depending on the alloy's composition, this temperature can range greatly, with some BMGs having crystallization temperatures up to 0.7–0.8 times their melting point. It is crucial to remember that processing variables like chilling rate and annealing temperature can also have an impact on the crystallization behavior of BMGs. These factors may have an impact on the development of crystalline stages and nucleation sites, which may alter the temperature at which crystals form.

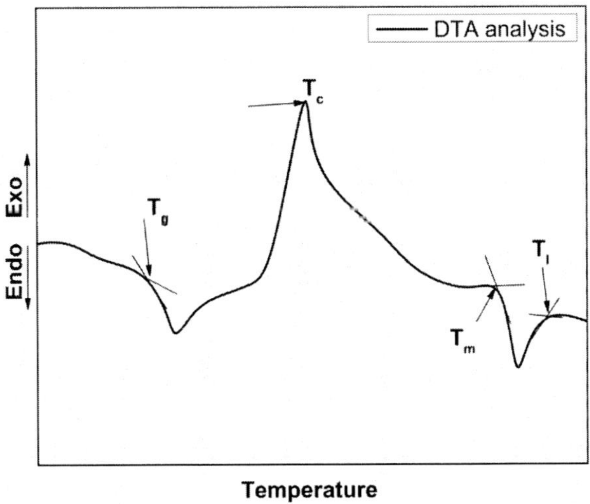

Figure 3. DTA Profile of BMGs and its parameters.

The $NMST_6$ sample gave the highest T_c value (969 K) and $NCSNb_6$ gives the least crystallization temperature (780 K) (Table 1 & Figure 4) (Kim et al., 2016; Y. Li et al., 2012; Samavatian, Gholamipour, Samavatian, & Farahani, 2019).

Supercooled Liquid

A key factor defining the thermal stability of metallic glasses is the super-cooled liquid (SLR) area. Its definition states that it is the differential in temperature between the glass transition temperature (T_g) and the temperature at which crystallization begins T_c. Alternatively put, the SLR is the range of temperatures where the metallic glass is still in a metastable supercooled liquid condition before it starts to crystallize. Depending on the alloy's composition and processing conditions, the SLR for metallic glasses can differ considerably. BMGs with greater glass-forming capacity typically have bigger SLRs. As an illustration, while some BMGs have SLRs of several hundred degrees Celsius, others only have SLRs of a few (Y. L. Wang & Xu, 2008).

Table 1. Thermal analysis for some BMG samples after 30 h of ballmilling

S.No	BMG	Composition	T_g (K)	T_c (K)	T_l (K)	$\Delta T = T_c - T_g$ (K)	$\Gamma = T_c/(T_g+T_l)$	$T_{rg} = T_g/T_l$	Ref.
1	$NMST_6$	$[NiMoSi]_{94}:[Ti]_6$	846	969	1267	123	0.458	0.667	Guddla et al., 2022
2	$NMSNb_{10}$	$[NiMoSi]_{90}:[Nb]_{10}$	815	976	1168	161	0.492	0.697	Guddla et al., 2021
3	$NCST_8$	$[Ni_{75}Cr_{15}Si_{10}]_{92}:[Ti]_8$	670	835	1330	165	0.417	0.503	Guddla et al., 2021b
4	$NCSNb_6$	$[NiCrSi]_{94}:[Nb]_6$	613	780	1226	167	0.424	0.500	Guddla et al., 2022b
5	$[TiZrCu]Ni_{10}$	$[Ti_{34}Zr_{75}Cu_{17.5}]:Ni_{10}$	698	841	1169	143	0.450	0.591	(M. Zhang, Song, Lin, Li, & Li, 2022)
6	MgNiNb	$Mg_{70}Ni_{15}Nd_{15}$	467	580	844	113	0.442	0.553	Huang et al., 2021
7	[LaAlNi]Cu	$[La_{55}Al_{25}Ni_{15}]Cu_5$	473	548	899	75	0.399	0.526	Yang et al., 2014

Table 2. Composition and Mechanical properties of ternary BMGs (From literature review) (GJ Hao et al., 2010)

Element			Composition, at. %			σ_y (MPa)	σ_f (MPa)	E (GPa)	HV
1	2	3	1	2	3				
Al	La	Cu	35	55	1		880	43	256
Cu	Hf	Ti	60	25	15	2010	2160	124	
Cu	Zr	Ag	50	45	5		1940	112	599
Cu	Zr	Al	47.5	47.5	5	1547	2265	87	
Cu	Zr	Al	46	46	8	1894	2250		580
Cu	Zr	Al	47	47	6	1733	2250		580
Cu	Zr	Ti	60	30	10	1785	2150	114	
La	Al	Ni	50	35	15	950	715	41	290
La	Al	Ni	55	25	20	735	515	34	225
Mg	Cu	Y	80	10	10	630	820		220
Mg	Ni	Y	82.5	12.5	5	610	44	212	
Zr	Al	Ni	70	10	20	1411	1335	61	432
Zr	Al	Ni	60	20	20	1795	1720	78.2	549
Zr	Ni	Al	60	25	15	1640	1715		502

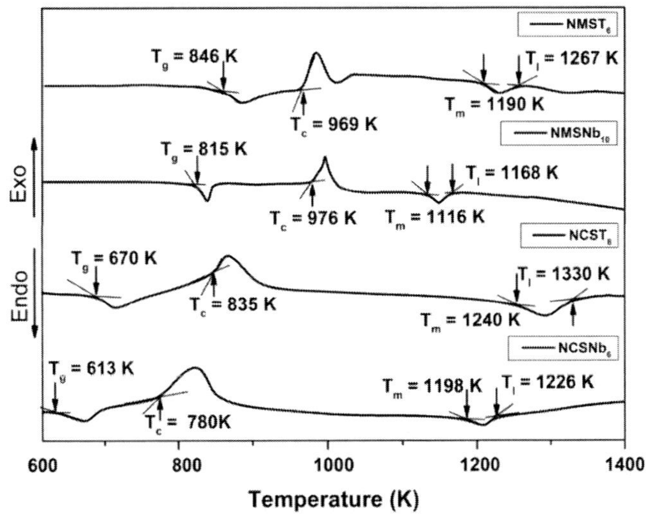

Figure 4. DTA profiles all as prepared $NMST_6$, $NMSNb_{10}$, $NCST_8$ and $NCSNb_6$ BMG samples at 30 h of ball milling.

For a better grasp of the thermal stability and processing of metallic glasses, the SLR is a crucial parameter. The metallic glass is frequently heated during processing, which can cause crystallization and damage to its mechanical qualities. For metallic glass, a bigger SLR can offer a larger processing window, enabling more adaptable processing conditions and enhanced mechanical properties. The SLR can also shed light on the fundamental physics of the stability and formation of metallic glasses. Investigations into the mechanisms that control the SLR continue to be an active area of study. ΔT (=167 K) is found to be the maximum for the sample, $NCSNb_6$ (Table 2). According to Turnbull, samples with a Trg value higher than 0.5 had better thermal stability and glass-forming capacity. As a result, we discover that every sample we're looking at has great amorphous properties. Table 1 makes it clear that the decreasing trend of the super-cooled liquid region width (ΔT) and the parameter is NCSNb6 > NCST8 > NMSNb10 > NMST6, which supports the claim that the NCSNb6 sample has a better glass-forming capacity than other BMGs. The compositional dependence of differential parameters of all synthesized NMST6, NMSNb10, NCST8, and NCSNb6 BMG samples show higher γ and T correlation with higher glass-forming capacity.

Mechanical Properties

Metallic glasses have distinctive qualities that include strength, hardness, wear resistance, and a good elastic limit. The plastic deformation is localized in the narrow shear bands, and does not exhibit strain hardening, and the plastic deformation is restricted to the small shear bands. This characteristic causes BMGs to have better compressive mechanical properties than tensile ones. Because of this characteristic, BMGs have superior compressive mechanical capabilities as opposed to tensile ones. BMGs' exceptional strength is a result of their crystalline structure and lack of dislocations. Mostly the strength varies as, Cu- based ~2GPa, Zr-, Ti- based ~3 GPa, Ni- based ~4GPa, Fe-based ~5GPa and Co- based ~6GPa (Louzguine Luzgin & Inoue, 2013; Louzguine Luzgin, Louzguina Luzgina, & Churyumov, 2012; Reddy, Bhattacharjee, & Murty, 2022; Y. Zhang, Yao, Zhao, & Ma, 2019).

Further improvements in understanding the GFA and the failure mechanisms of BMGs, paved the way for creating a wide range of alloy compositions, and improvement of strength, ductility, and fracture toughness needed for advanced structural applications. These unique mechanical properties of BMGs combined with the ease of fabrication into bulk forms draw attraction towards aerospace, biomedical, electronic, sports goods, luxury goods, armor, and naval applications. The mechanical properties such as Yield strength ($\boldsymbol{\sigma_y}$), Fracture strength ($\boldsymbol{\sigma_f}$), Young's Modulus (**E**), and

Vickers Hardness (**Hv**) values of certain combinations of ternary and quaternary BMGs are shown in Table 2 and Table 3 (GJ Hao, Lin, Zhang, Chen, & Lu, 2010).

Table 3. Composition and Mechanical properties of quaternary BMGs (From literature review) (GJ Hao et al., 2010)

Element				Composition, at. %				σ_y (MPa)	σ_f (MPa)	E (GPa)	HV
1	2	3	4	1	2	3	4				
Ce	Al	Cu	Co	68	10	20	2		1180	31.34	
Cu	Hf	Ti	Ta	56.4	23.5	14.1	6	2125	2100	104	
Cu	Zr	Ag	Al	45	45	7	3		1836	110	540
Cu	Zr	Ti	Y	58.8	29.4	9.8	2	1780	2050	115	
La	Al	Cu	Ag	55	15	20	10		758	42	208
Ni	Si	B	Nb	72	7.68	16.32	4		2510	77	870
Ni	Si	B	Ta	72	7.68	16.32	4		2730	75	920
Pd	Pt	Cu	P	35	15	30	20	1410			470
Zr	Al	Co	Cu	55	20	20	5	2000	1960	92	
Zr	Cu	Ni	Al	52	32	4	12		1780	88	501
Zr	Cu	Ni	Al	52	26	10	12		1960	89	509
Zr	Fe	Al	Cu	60	10	7.5	22.5	1718		100	

Compression Test

Apart from the tensile test and flexion test, the compression test is the most fundamental type of mechanical testing. Uniaxial compression test frequently produces larger strains without failure of the specimen, mostly used to determine the material behavior under applied crushing loads. During the test process, the various properties of the material under investigation, are calculated and plotted as a stress-strain diagram (Figure 5), which is used to calculate the yield strength, elastic limit, yield point, compressive strength, and proportional limit of the material.

The crystalline lattice and lack of dislocations are the causes of the high strength of BMGs. Mostly, compression tests are done on brittle materials such as metals, plastics, concrete, ceramics, composites, and cardboard. More often, compression tests are performed on finished products like protective cases, golf balls, tennis balls, water bottles, plastic pipes, and furniture. An engineer has to balance between material conservation and also product strength, for example, he may want to conserve plastic by making bottles with thinner walls, and at the same time, the bottles should be strong enough to be stacked up during transport without breaking (Chang, Wang, Ge, Zhou, & Cui, 2018; Egami, Iwashita, & Dmowski, 2013; T. Zhang & Inoue, 2002).

Table 4. Mechanical Properties of all prepared BMG samples after 30 h of ball milling

BMG	Composition	Yield Strength, σ_y (MPa)	Fracture Strength, σ_f (MPa)	Plastic Strain, ε_p (%)	Hardness (Hv)	Ref
$NMST_6$	$[NiMoSi]_{94}:[Ti]_6$	1361 ± 47	1451 ± 39	0.92 ± 0.1	650	Guddla et al.,2022
$NMSNb_{10}$	$[NiMoSi]_{90}:[Nb]_{10}$	1262 ± 21	1378 ± 12	0.661 ± 0.1	575	Guddla et al., 2021
$NCST_8$	$[Ni_{75}Cr_{15}Si_{10}]_{92}:[Ti]_8$	1241 ± 32	1326 ± 23	0.68 ± 0.1	598	Guddla et al., 2021b
$NCSNb_6$	$[NiCrSi]_{94}:[Nb]_6$	1363 ± 43	1467 ± 45	0.947 ± 0.1	673	Guddla et al., 2022b

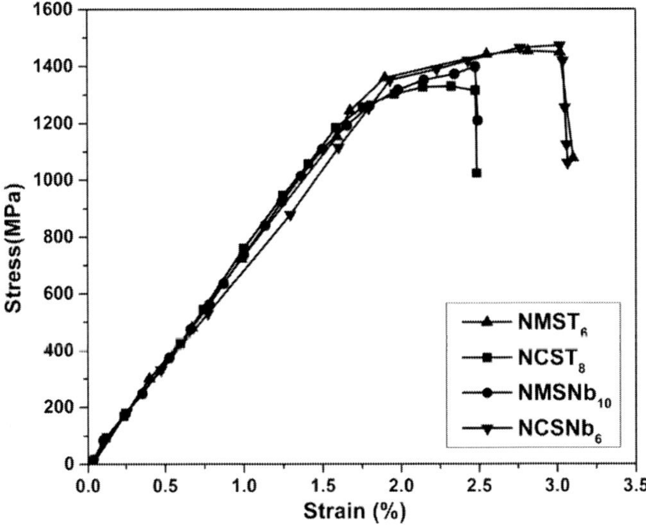

Figure 5. Stress-strain curves of all as-prepared $NMST_6$, $NMSNb_{10}$, $NCST_8$, and $NCSNb_6$ BMG samples.

Table 4 shows all the calculated values of mechanical properties of all as-prepared BMG samples after 30 h of ball milling. It is seen that $NCSNb_6$ samples have the maximum values of yield strength (σ_y), fracture strength (σ_f), plastic strain (ε_p), and hardness (Hv) values.

Vickers Hardness Testing Machine

The hardness of a material is defined as the quality to withstand localized deformation. Soft materials usually suffer indentations whereas hard materials can withstand to any change in shape. This property is important especially when we look for a suitable material in an environment that includes tiny particles that can induce wear. The various types of hardness include:

- Indentation hardness
- Brinell Hardness
- Rockwell Hardness
- Scratch Hardness
- Vickers Hardness
- Mohs Hardness
- Scleroscope Hardness

Tungsten (1960–2450 MPa), Iridium (1670 MPa), Steel, Osmium (3920–4000 MPa), Chromium (687–6500 MPa), and Titanium (716–2770 MPa) are discovered to be the world's hardest metals in the world(Pan et al., 2005; Z. Zhang, Eckert, & Schultz, 2003). Vickers Hardness Test is an optical technique that uses the indentation size (left by the indenter) to determine the material's hardness. It is a static harness testing technique, and the softer the material is, the more indent is left. According to ISO 6507, a diamond pyramid-shaped indenter is pressed into the specimen at a test load of 1 kgf and an interfacial angle of 1360 to measure the Vickers Hardness. Vickers Pyramid number (HV) or Diamond Pyramid Hardness (DPH) are the units of hardness. The formula HV = 1.854 (F/D2) can be used to determine the Vickers Hardness number (HV), where "F" stands for the applied load in kgf and "D2" for the area of indentation in mm^2. Universal Wolpert Wilson 930/250 N The user-friendly operator screen of the DigiTestor Vickers indenter displays the hardness results of all test processes and has a range of 1–250 Kgf. It offers a simple user interface with a one-touch panel that delivers effective and trustworthy test results for test sizes between 20X and 140X. Vickers hardness and plastic strain values for all manufactured $NMST_6$, $NMSNb_{10}$, $NCST_8$, and $NCSNb_6$ BMG samples are correlated in Figure 6.

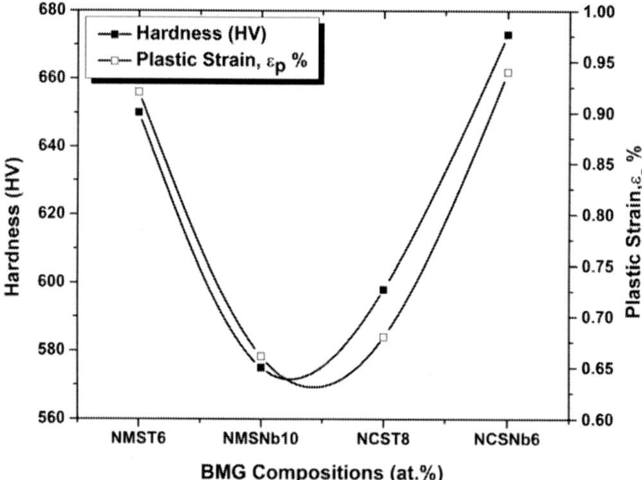

Figure 6. Comparison for Vickers hardness and plastic strain values for certain BMG based composition.

Applications of Vickers Hardness

- Find the hardness of castings and forgings.
- Space and Automotive industry.
- Can be used for both flat and cylindrical work pieces.
- Quality control tests or Sample testing.
- Can be used with steels and non-ferrous materials.
- Ceramics, stainless steel and cemented carbides.
- Used in Laboratories and workshops, etc.

Advantages of Vickers Hardness

- This procedure contains an entire hardness range from all soft to hard materials.
- Since this method belongs to Non-destructive testing method, the sample can be used for any other purposes.
- Only one type of indenter is used for all types of Vickers methods.

Dis-Advantages of Vickers Hardness

- This process is rather slow as compared to other methods.
- The specimen to be tested should have good surface quality as the indent is measured optically.
- As it uses an optical system, this equipment is costlier than Rockwell tester.

Corrosion Properties

Introduction to Corrosion

Corrosion is a natural process that occurs when metals are exposed to environmental factors such as moisture, oxygen, and chemicals. The corrosion process can lead to damage, loss of material, and reduced structural integrity of the metal. Corrosion properties refer to the behavior of a metal when it is exposed to different environments and how it is affected by the corrosion process. Corrosion is a natural process that occurs when materials such as metals, plastics, and ceramics are exposed to their environment. It is a gradual degradation of a material by chemical or electrochemical reaction with its surroundings, which causes changes in the properties of the material, including its strength, appearance, and other physical and mechanical characteristics (Debnath, Kim, & Fleury, 2012; C. Qin et al., 2006; F. X. Qin et al., 2016; F. Qin et al., 2007; Qiu, Liu, Sun, & Zhang, 2005; Scully, Gebert, & Payer, 2007).

Corrosion is a widespread and expensive problem that affects almost every aspect of our daily lives, from the infrastructure we rely on, such as bridges and buildings, to the vehicles we use to get around. The effects of corrosion can range from minor aesthetic damage to catastrophic failure of critical components, leading to safety hazards, environmental damage, and economic losses. The numerous types of corrosion:

a. Stress corrosion cracking
b. Uniformcorrosion,
c. Pittingcorrosion,
d. Galvaniccorrosion,
e. Crevicecorrosion.

The sort of corrosion that develops is influenced by the material, the surrounding environment, and the usage circumstances. Engineers and scientists have a huge problem in preventing or regulating corrosion, and numerous methods have been developed to lessen its effects. These methods include the use of corrosion-resistant materials, protective coatings, cathodic protection, and corrosion inhibitors (An et al., 2013; Asami, Qin, Zhang, & Inoue, 2004; Azadmanjiri et al., 2018; Cai et al., 2012; Cottis, 2010; Tsai et al., 2020).

Wet and dry corrosion, low- and high-temperature corrosion, and other classifications are all possible for corrosion. Aqueous solutions are where wet corrosion happens, whereas high temperatures and the absence of a liquid medium are where dry corrosion usually takes place. The electrochemical process, which involves anodic (oxidation) and cathodic (reduction) reactions at the interface between the metal surface and an electrolyte, causes corrosion in aqueous environments. An oxidation process takes place at anodic sites, which results in the loss of electrons as:

$$M \rightarrow M^{n+} + ne^{-} \quad 2.1$$

At the same time, the cathodic sites undergo a reduction process, which is the recovery of electrons as:

$M'^{n+} + ne^- \rightarrow M'$ 2.2

Anodic and cathodic reactions proceed at the same pace because of the charge balance. Heterogeneity (such as differences in composition or grain size, surface roughness, impurities or inclusions, and localized stresses) or between two different metals exposed to a corrosive environment can cause anodic and cathodic sites to form on the surface of a metal. Anodic reaction, cathodic reaction, the presence of an electrolyte, and electrical connectivity are the four main requirements for electrochemical corrosion.

Important Corrosion Properties of Metals

The most frequent type of corrosion, uniform corrosion happens uniformly over a material's whole surface and causes a general thinning of the material. Normally, electrochemical reactions between the material and its surroundings, including exposure to moisture, oxygen, or corrosive chemicals, are what trigger it. The chemical makeup and physical characteristics of the material, the type and intensity of the corrosive environment, temperature, and additional environmental elements including humidity, pH, and salinity all have an impact on the rate of uniform corrosion (H. Lu, Zhang, Gebert, & Schultz, 2008; Naka, Hashimoto, & Masumoto, 1976). Figure 7 shows the types of corrosion as discussed below in detail.

Uniform Corrosion

Uniform corrosion can lead to a reduction in the thickness of the material, which can weaken its mechanical strength and cause failure or collapse. This type of corrosion can occur in various materials, including metals, plastics, and ceramics, and it is a significant problem in industries such as manufacturing, transportation, and construction. Several techniques can be used to prevent or mitigate uniform corrosion, including the use of corrosion-resistant materials, protective coatings, and inhibitors. Regular maintenance and inspections can also help detect and address early signs of uniform corrosion before it causes significant damage.

Pitting Corrosion

The resistance of a material to localized corrosion or pitting corrosion is measured by the amount of corrosion that it experiences. Pitting corrosion is a particular kind of corrosion that takes place if there is a small crack or flaw in the passivation layer or protective coating, allowing aggressive ions to attack the underlying material. Pitting corrosion can cause the material's surface to develop microscopic pits or craters, which can reduce the material's mechanical strength and cause early failure. Pitting corrosion is a typical issue in situations with high amounts of chloride or other aggressive ions and can affect a variety of materials, including metals, polymers, and ceramics (Niu et al., 2020).

The chemical makeup, microstructure, type, and concentration of the corrosive environment all affect a material's ability to withstand pitting. Higher concentrations of corrosion-resistant alloying elements, such as chromium, molybdenum, and nickel, which create a passive layer of protection on the surface of the material, are often found in materials that are more resistant to pitting corrosion. A variety of tests, such as the ASTM G48 test, which evaluates a material's critical pitting temperature (CPT) and critical pitting index (CPI), can be used to assess a material's ability to resist pitting. The results of these tests can be used to select materials with the appropriate pitting resistance for specific applications, such as in marine or chemical processing industries, where pitting corrosion is a significant concern.

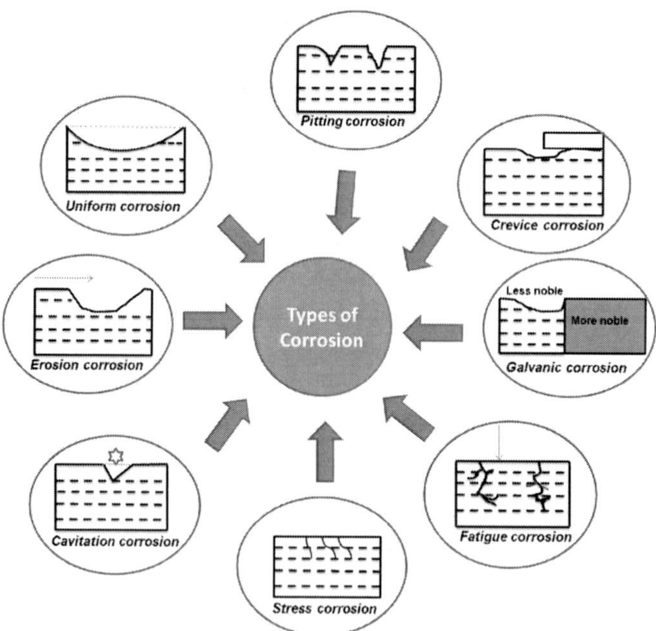

Figure 7. Types of corrosion in BMGs.

Crevice Corrosion

Crevice corrosion is a type of localized corrosion that occurs in crevices or narrow gaps between two surfaces that are in contact with each other. It is caused by the accumulation of corrosive substances, such as salt, moisture, or acids, in the crevice, which creates an environment that is more aggressive than the surrounding environment. Crevice corrosion can occur in various materials, including metals, plastics, and ceramics, and is commonly found in joints, gaskets, seals, and fasteners. It can lead to the formation of deep pits or cracks, which can weaken the mechanical strength of the material and cause premature failure.

The severity of crevice corrosion depends on several factors, including the geometry and size of the crevice, the nature and concentration of the corrosive substances, the temperature, and the duration of exposure. Materials that are more susceptible to crevice corrosion include those with lower corrosion resistance or those that are more likely to form a stable oxide layer on the surface, such as stainless steel. Preventing or mitigating crevice corrosion requires careful design and material selection, as well as regular inspections and maintenance. Techniques such as minimizing crevice geometry, using corrosion-resistant materials, applying coatings or sealants, and providing adequate ventilation or drainage can help reduce the risk of crevice corrosion. In some cases, cathodic protection or the use of corrosion inhibitors may also be effective in controlling crevice corrosion (C. Zhang et al., 2023).

Galvanic Corrosion

When two distinct metals or alloys come into touch with one another while in the presence of an electrolyte, such as water or a corrosive solution, galvanic corrosion, a kind of electrochemical corrosion, happens. Here, one metal serves as an anode and corrodes, while the other metal serves as a cathode and is shielded. The nature and composition of the metals or alloys, their surface area, the type and concentration of the electrolyte, and the temperature all affect how quickly galvanic corrosion occurs. The likelihood of galvanic corrosion increases with the degree of difference between the two metals or alloys. Galvanic corrosion can lead to the formation of pits or holes on the surface of the metal, which can weaken its mechanical strength and lead to premature failure. It is a common problem in various industries, including marine, construction, and electronics, where dissimilar metals or alloys are often used in close proximity. Several techniques can be used to prevent or mitigate galvanic corrosion, including the use of coatings, inhibitors, and sacrificial anodes. Coatings, such as paints or plating's, can be applied to the surface of the metal to prevent contact with the electrolyte. Inhibitors can be added to the electrolyte to reduce the rate of corrosion, while sacrificial anodes, such as zinc or magnesium, can be used to protect the metal by corroding in place of the more valuable metal. Designers and engineers must consider the potential for galvanic corrosion when selecting materials and designing structures to ensure the longevity and safety of the components. Regular inspections and maintenance can also help detect and address early signs of galvanic corrosion before it causes significant damage (Koga et al., 2021).

Erosion Corrosion

Erosion corrosion, also known as abrasive corrosion, is a type of corrosion that occurs when a metal surface is exposed to a fluid or gas that contains abrasive particles or when the fluid flow causes a relative motion between the surface and the fluid. The abrasion causes the metal to wear away faster than it would under normal circumstances, leading to corrosion. Erosion corrosion is often seen in pipes and other fluid-carrying systems, especially those carrying high-velocity fluids or fluids containing abrasive particles such as sand or grit. It can also occur in machinery parts that are exposed to fluids with high velocity or turbulence, such as pump impellers, valve bodies, and propellers.

To prevent erosion-corrosion, engineers may choose to use materials that are more resistant to abrasion, such as stainless steel or alloys that are specifically designed for high-velocity applications. Other methods of prevention include reducing fluid velocity, minimizing turbulence, and filtering out abrasive particles from the fluid. Regular inspections and maintenance are important in detecting and preventing erosion-corrosion, as it can lead to leaks and failures in pipes and machinery if left unchecked (Ji, Zhao, Zhang, & Zhou, 2013).

Cavitation Corrosion

Cavitation corrosion can be a particular problem BMGs because these materials have a unique atomic structure that makes them highly susceptible to localized deformation and cracking. This susceptibility can be exacerbated by the mechanical stresses that are generated during cavitation, which can cause significant damage to the BMG surface. One approach to mitigating cavitation corrosion in BMGs is to improve their mechanical properties through the use of alloying elements and/or heat treatments. By enhancing the strength and ductility of the BMG, it may be possible to reduce the likelihood of cavitation-induced cracking and other forms of damage (Drozdz, Wunderlich, & Fecht, 2007).

Another approach is to modify the surface properties of the BMG through the use of coatings or other surface treatments. These treatments can improve the material's resistance to cavitation by reducing the amount of energy that is absorbed by the surface during bubble collapse. In addition to these strategies, it is important to carefully design and operate equipment that utilizes BMGs in order to minimize the risk of cavitation corrosion. This may involve optimizing the geometry and flow conditions of pumps and other machinery to reduce the likelihood of cavitation, as well as implementing regular maintenance and inspection programs to detect and address any issues that do arise.

Stress Corrosion

Stress corrosion cracking (SCC) is a certain type of corrosion that occurs under conditions of tensile stress and exposure to a corrosive environment, leading todefects that can act as initiation sites for cracking. SCC can be a serious issue for BMGs, as these materials are highly susceptible to cracking under the right conditions. One potential cause of SCC in BMGs is the presence of hydrogen in the environment. Hydrogen can diffuse into the material and cause embrittlement, reducing the material's ability to resist tensile stresses and increasing the likelihood of cracking. Another potential cause of SCC in BMGs is the formation of microcracks or other defects in the material. To prevent SCC in BMGs, it is important to carefully control the material's environment and operating conditions. This may involve using corrosion-resistant coatings or other protective measures to reduce the material's exposure to corrosive agents, as well as minimizing the material's exposure to tensile stresses. In addition, it is important to carefully monitor the material's condition over time, using non-destructive testing techniques to detect any signs of cracking or other forms of damage. Early detection and intervention can help to prevent catastrophic failure and ensure the long-term reliability of BMG components (Nakai & Yoshioka, 2010).

Fatigue Corrosion

Fatigue corrosion is a type of material degradation that occurs in metals when they are subjected to cyclic loading in a corrosive environment. This can be a particular problem in BMGs, which are highly susceptible to fatigue failure under certain conditions. The mechanisms that lead to fatigue corrosion in BMGs are complex and not yet fully understood. However, it is believed that the repeated loading and unloading of the material can cause localized deformation and stress concentrations, which can lead to the initiation and propagation of cracks. In the presence of a corrosive environment, these cracks can grow more quickly and eventually lead to component failure (Peter et al., 2002).

To prevent fatigue corrosion in BMGs, it is important to carefully control the material's environment and operating conditions. This may involve using corrosion-resistant coatings or other protective measures to reduce the material's exposure to corrosive agents, as well as minimizing the material's exposure to cyclic loading. In addition, it is important to carefully monitor the material's condition over time, using non-destructive testing techniques to detect any signs of cracking or other forms of damage. Regular inspection and maintenance can help to detect any developing issues before they become more serious. Designing components that are less susceptible to fatigue corrosion can also be helpful. This may involve optimizing the geometry and load conditions of components, as well as using materials that have been specifically engineered to resist fatigue failure in corrosive environments.

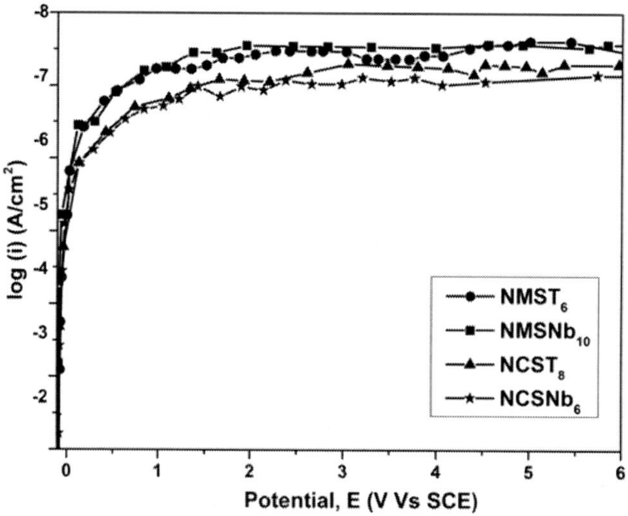

Figure 8. Potentiometric analysis of $NMST_6$, $NMSNb_{10}$, $NCST_8$ and $NCSNb_6$ BMG samples at 30 h of ball milling - Anodic current density with corrosion potential.

Table 5. Variation potential and current parameters for different $NMST_6$, $NMSNb_{10}$, $NCST_8$, and $NCSNb_6$ BMG samples

BMG	Composition	E_c(mV)	E_p (V)	i_c (A/cm^2)	i_p(A/cm^2)	Ref
NMST6	$[NiMoSi]_{94}:[Ti]_6$	-71.2	0.521	4.7 x10-4	0.0031	Guddla et al., 2022
$NMSNb_{10}$	$[NiMoSi]_{90}:[Nb]_{10}$	-61.2	0.51	4.1 x10-4	0.0066	Guddla et al., 2021
$NCST_8$	$[Ni_{75}Cr_{15}Si10]_{92}:[Ti]_8$	-68.5	0.5423	4.7 x10-4	0.0079	Guddla et al., 2021b
$NCSNb_6$	$[NiCrSi]_{94}:[Nb]_6$	-65.4	0.5120	5.2 x10-3	0.0062	Guddla et al., 2022b

Figure 8 shows the potentiometric analysis of $NMST_6$, $NMSNb_{10}$, $NCST_8$, and $NCSNb_6$ BMG samples prepared at 30 h of the ball milling method. Table 5 shows the potentiometric analysis for different properties such as corrosion potential (E_c), passive potential (E_p), cathode current density (i_c), and anode current density (i_p) for prepared $NMST_6$, $NMSNb_{10}$, $NCST_8$, and $NCSNb_6$ BMG samples. It is seen that the $NCSNb_6$ sample had the least values of i_c (= 5.2X10^{-3} A/cm^2) and i_p (= 0.0062 A/cm^2) values.

Applications of BMGs

Nowadays BMGs are a hot topic among scientific researchers owing to their use in many novel commercial and potential applications including aerospace (Egami, 2010; Inoue & Takeuchi, 2011; Khan et al., 2018; Kruzic, 2016; Nishiyama, Amiya, & Inoue, 2007; Nu & Luong, 2016; Patel, Swain, Behera, & Mohapatra, 2020)[Nu and Luong,2016; Patel et al., 2020; Inoue and Takeuchi2011; Nishiyama et al., 2007; Shen and Inoue, 2003; Khan et al., 2018; Egami, 2010;Kruzic, 2016]. To sum up few, here are some applications of BMGs as follows:

- The commercial BMG composition [Zr-Ti-Cu-Ni-Be], developed for use in aerospace materials.
- [Ti-Zr-Cu-Ni-Sn]- The Coriolis flow meter, which is 28–53 times more sensitive than a typical flow meter, uses this BMG composition to increase sensitivity.
- In comparison to typical stainless-steel sensors used in the automotive industry, sensors manufactured from [Zr-Al-Ni-Cu] BMG are found to be smaller, more sensitive, and more pressure-resistant. Additionally, the world's smallest geared motor, with a 1.5 mm and 9.9 mm diameter, is made from this alloy.
- The special magnetic properties of ferromagnetic BMGs are used in high-efficiency transformers.
- The unique softening behavior of the BMGs is used in thermoplastics.
- The nano-moulds made from BMGs are more durable and easier to fabricate than the silicon molds used in nano-imprint lithography.
- They are used in electronic applications due to their superior electronic, mechanical, and thermal properties.

- [Ti-Cu-Pd-Zr] BMG composition is known to be non-carcinogenic and is used in bio-mechanical applications, allowing better joining with bones using laser pulses, as it is 3 times stronger than Ti and elastic modulus nearly matching with bones.
- Helpful in making biomedical implants owing to their lightweight and being tougher than steel and titanium counterparts. Metallic glass made of Mg-Zn-Ca is both biocompatible and biodegradable.
- [Mg-Zn-Ca] composition is used as a bio-material to be replaced with bone tissue as screws, plates, and pins in order to fix fractures.
- Used in golf clubs for making striking face plates, owing to the low density, higher strength-to-weight ratio, and lower elastic modulus. It is said from literature that less energy is absorbed at the impact point of the club head and more energy is transmitted to the ball. A steel club transmits 60%, a titanium club transmits 70%, and a metallic glass transmits 99% of the input energy.
- Used in making frames in tennis rackets.
- Used in making mobile phones and digital still cameras - Cellular phone casing.
- Used in optical fibre connections.
- Used in making shot-peening balls.
- Used in high-frequency magnetic coils and also as tiny compasses.
- Used in lead-free soldering in vessels.
- Used in high-corrosion-resistant coating for plates, etc.
- Other potential applications include fishing equipment, sporting goods, guns, hunting bows, scuba gear, bicycle frames, and marine applications.
- Used in US Army, for the development of metallic glass tank armor penetrator bullets, in place of uranium penetrators to avoid biological toxicity.
- Used in making tools like knives and razors and scalpel blades that are less expensive than diamond and last longer and sharper than steel.

Conclusion

We studied all the topics on Bulk Metallic Glasses in detail focussing on synthesis using ball milling apparatus and characterization tests based on thermal, mechanical, and corrosion properties. Also discussed various applications of BMGs. Optimum thermal (ΔT, $T_{rg}$and γ), mechanical (σ_y, ε_p, hardness), and potentiometric parameters (E_c, i_c, E_p and i_p) can be achieved for $[Ni\text{-}Cr\text{-}Si]_{100-x}:[Nb]_x$ prepared using Ball milling method and among all the synthesized BMGs in the chapter, $NCSNb_6$ based BMGs network is promising for lightweight applications. In addition to $NCSNb_6$, BMGs based on $[Ni\text{-}(Mo/Cr)\text{-}Si]_{100-x}:[Ti/Nb]_x$ also show optimum properties well suited for discussed applications.

References

Asami K, Qin CL, Zhang T, Inoue A. Effect of additional elements on the corrosion behavior of a c Cu-Zr-Ti bulk metallic glass. *Materials Science and Engineering: A* (2004) 375:235-239.

Azadmanjiri J, Wong WH, Kanwar JR, Berndt CC, Wang J, Srivastava VK, Kapoor A. (2018). Nanocoutured metallic biomaterials and surface functionalization of titanium-based alloys for medical applications. *Nanobiomaterials*: CRC Press 17-50.

Bhadeshia H. Thermal analyses techniques- Differential thermal analysis. *Material Science and Metallurgy* (2002).

Brink T, Adjaoud O, Albe K. Nanostructured metallic glasses: Tailoring the mechanical properties of amorphous metals. *NIC Symposium* (2016) 48:191-198.

Busch R. The thermophysical properties of bulk metallic glass-forming liquids. *Jom* (2000) 52:39-42.

Cai A, Xiong X, Liu Y, An W, Zhou G, Luo Y, Li T. Corrosion behavior of $Cu_{55}Zr_{35}Ti_{10}$ metallic glass in the chloride media. *Materials Chemistry and Physics* (2012) 134(2-3):938-944.

Cao D, Wu Y, Liu X, Wang H, Wang X, Lu Z. Enhancement of glass-forming ability and plasticity via alloying the elements having positive heat of mixing with cu in $Cu_{48}Zr_{48}Al_4$ bulk metallic glass. *Journal of Alloys and Compounds,* (2019) 777:382-391.

Cao D, Wu Y, Wang H, Liu XJ, Lu Z. Effects of nitrogen on the glass formation and mechanical properties of a ti-based metallic glass. *Acta Metallurgica Sinica (English Letters)* (2016) 29:173-180.

Cao G, Liu K, Liu G, Zong H, Bala H, Zhang B. Improving the glass-forming ability and the plasticity of Zr-Cu-Al bulk metallic glass by addition of nb. *Journal of Non-Crystalline Solids,* (2019) 513:105-110.

Chang Z, Wang W, Ge Y, Zhou J, Cui Z. Microstructure and mechanical properties of Ni-Cr-Si-B-Fe composite coating fabricated through laser additive manufacturing. *Journal of Alloys and Compounds* (2018) 747:401-407.

Chen CL, Huang CL. Milling media and alloying effects on synthesis and characteristics of mechanically alloyed ods heavy tungsten alloys. *International Journal of Refractory Metals and Hard Materials* (2014) 44:19-26.

Chen M. A brief overview of bulk metallic glasses. *NPG Asia Materials* (2011) *3*(9):82-90.

Chen N, Martin L, Luzguine Luzgin DV, Inoue A. Role of alloying additions in glass formation and properties of bulk metallic glasses. *Materials* (2010) 3(12):5320-5339.

Cheng Y, Sheng H, Ma E. Relationship between structure, dynamics, and mechanical properties in metallic glass-forming alloys. *Physical Review B* (2008) 78(1):014207.

Churyumov AY, Bazlov A, Tsarkov A, Solonin A, Louzguine Luzgin D. Microstructure, mechanical properties, and crystallization behavior of Zr-based bulk metallic glasses prepared under a low vacuum. *Journal of Alloys and Compounds* (2016) 654:87-94.

Cottis R. *Shreir's corrosion*. Amsterdam, The Netherlands: Elsevier; 2010.

Debenedetti PG, Stillinger FH. Supercooled liquids and the glass transition. *Nature* (2001) 410(6825):259-267.

Debnath MR, Kim DH, Fleury E. Dependency of the corrosion properties of in-situ Ti-based BMG matrix composites with the volume fraction of crystalline phase. *Intermetallics* (2012) 22:255-259.

Drozdz D, Wunderlich R, Fecht HJ. Cavitation erosion behaviour of Zr-based bulk metallic glasses. *Wear* (2007) 262(1-2):176-183.

Du X, Huang J, Liu C, Lu Z. New criterion of glass forming ability for bulk metallic glasses. *American Institute of Physics* (2007).

Egami T. Understanding the properties and structure of metallic glasses at the atomic level. *JOM,* (2010) 62(2):70-75.

Egami T, Iwashita T, Dmowski W. Mechanical properties of metallic glasses. *Metals* (2013) 3(1):77-113.

Feng Y, Cai A, Ding D, Liu Y, Wu H, An Q, Peng Y. Composition design and properties of Cu-Zr-Ti bulk metallic glass composites. *Materials Chemistry and Physics* (2019) 232, 452-459.

Filipecka K, Pawlik P, Filipecki J. The effect of annealing on magnetic properties, phase structure and evolution of free volumes in Pr-Fe-Bw metallic glasses. *Journal of Alloys and Compounds,* (2017) 694:228-234.

Gandi S, Chinta SR, Ghoshal P, Ravuri BR. SnO-GeO_2-Sb_2O_3 glass anode network mixed with different Ba^{2+} fractions: Investigations on na-ion storage capacity and stability. *Journal of Non-Crystalline Solids* (2019) 506:80-87.

Gu J, Yang X, Zhang A, Shao Y, Zhao S, Yao K. Centimeter-sized ti-rich bulk metallic glasses with superior specific strength and corrosion resistance. *Journal of non-crystalline solids* (2019) 512:206-210.

Gu X, Poon SJ, Shiflet GJ. Mechanical properties of iron-based bulk metallic glasses. *Journal of Materials Research* (2007) 22(2):344-351.

Guddla G, Katta V, Gandi S, Ambadipudi S, Ravuri B. The role of titanium content in $(Ni_{75}Cr_{15}Si_{10})_{100-x}Ti_x$ bulk metallic glass systems to elevate mechanical and corrosion properties. *Phase Transitions* (2021) 94(10):679-690.

Guddla GT, Ambadipudi S, Katta VK, Katari NK, Ravuri BR. Influence of titanium content on thermal, mechanical and corrosion behaviour anomalies of nickel-molybdenum-silicate bulk metallic glasses. *Silicon* (2022) 14(4):1571-1581.

Guddla GT, Ambadipudi S, Yenduva S, Katta VK, Ravuri BR. [Ni-Mo-Si]: Nb bulk metallic glasses: Microstructure, mechanical and corrosion studies. *Silicon* (2022) 14(6):2545-2553.

Guddla GT, Gandi S, Ambadipudi S, Ravuri BR. Design of Ni-based bulk metallic glasses with improved mechanical and corrosion properties. *Current Applied Science and Technology* (2021) 21:115-131.

Guo S, Liu Z, Chan KC, Chen W, Zhang H, Wang J, Yu P. A plastic ni-free Zr-based bulk metallic glass with high specific strength and good corrosion properties in simulated body fluid. *Materials Letters* (2012) 84:81-84.

Hao G, Lin J, Zhang YJ, Chen G, Lu Z. Ti–Zr–Be ternary bulk metallic glasses correlated with binary eutectic clusters. *Materials Science and Engineering: A* (2010) 527(23):6248-6250.

Hao G, Ren F, Zhang Y, Lin J. Role of yttrium in glass formation of Ti-based bulk metallic glasses. *Rare Metals* (2009) 28:68-71.

Hasegawa M, Takeuchi M, Kato H, Inoue A. Effects of a small amount of si or Ge addition on stability and hydrogen-induced internal friction of $Ti_{34}Zr_{11}Cu_{47}Ni_8$ glassy alloys. *Acta materialia* (2004) 52(7):1799-1806.

He Q, Shang JK, Ma E, Xu J. Crack-resistance curve of a Zr-Ti-Cu-Al bulk metallic glass with extraordinary fracture toughness. *Acta materialia* (2012) 60(12):4940-4949.

Hitit A, Şahin H, Öztürk P, Aşgın AM. A new ni-based metallic glass with high thermal stability and hardness. *Metals* (2015) 5(1):162-171.

Hofmann DC. Bulk metallic glasses and their composites: A brief history of diverging fields. *Journal of Materials* (2013) 2013:517904.

Inoue A. *Bulk glassy alloys, practical characteristics and applications*. Zurich: Trans Tech Publications; 1999.

Inoue A. Stabilization of metallic supercooled liquid and bulk amorphous alloys. *Acta materialia* (2000) 48(1):279-306.

Inoue A, Kong F. Bulk metallic glasses: Formation and applications, *Encyclopedia of Materials: Science and Technology*, Amsterdam, Netherlands: Elsevier; 2016.

Inoue A, Kong F, Zhu S, Al Marzouki F. Peculiarities and usefulness of multicomponent bulk metallic alloys. *Journal of Alloys and Compounds* (2017)707:12-19.

Inoue A, Shen B, Nishiyama N. Development and applications of late transition metal bulk metallic glasses. *Bulk metallic glasses,* Boston, MA: Springer; 2008.

Inoue A, Takeuchi A. Recent development and application products of bulk glassy alloys. *Acta Materialia* (2011) 59(6):2243-2267.

Inoue A, Zhang T, Masumoto T. Zr-Al-Ni amorphous alloys with high glass transition temperature and significant supercooled liquid region. *Materials Transactions, JIM* (1990) 31(3):177-183.

Inoue A, Zhang T, Masumoto T. Glass-forming ability of alloys. *Journal of non-crystalline Solids* (1993) 156:473-480.

Ji X, Zhao J, Zhang X, Zhou M. Erosion–corrosion behavior of zr-based bulk metallic glass in saline-sand slurry. *Tribology International* (2013) 60:19-24.

Katta VK, Gandi S, Katari NK, Mekprasart W, Pecharapa W, Dutta DP, Ravuri BR. Mixed polyanion Na-Mn-V-P glass–ceramic cathode network: Improved electrochemical performance and stability. *Energy Technology* (2021) 9(2):2000845.

Kawamura Y, Nakamura T, Inoue A. Superplasticity in $Pd_{40}Ni_{40}P_{20}$ metallic glass. *Scripta Materialia* (1998) 39(3):301-306.

Khalifa HE. Bulk metallic glasses and their composites: Composition optimization, thermal stability, and microstructural tunability. PhD Thesis, San Diego: University of California; 2009.

Khan MM, Nemati A, Rahman ZU, Shah UH, Asgar H, Haider W. Recent advancements in bulk metallic glasses and their applications: A review. *Critical Reviews in Solid State and Materials Sciences* (2018) 43(3):233-268.

Kim J, Kyeong JS, Ham MH, Minor AM, Kim DH, Park ES. Development of Mo-Ni-Si-B metallic glass with high thermal stability and h versus e ratios. *Materials & Design* (2016) 98:31-40.

Klement W, Willens R, Duwez P. Non-crystalline structure in solidified gold–silicon alloys. *Nature,* (1960) 187(4740):869-870.

Koga G, Ferreira T, Guo Y, Coimbrão D, Jorge Jr A, Kiminami C, Botta W. Challenges in optimizing the resistance to corrosion and wear of amorphous Fe-Cr-Nb-B alloy containing crystalline phases. *Journal of Non-Crystalline Solids* (2021) 555:120537.

Kruzic JJ. Bulk metallic glasses as structural materials: A review. *Advanced Engineering Materials* (2016) 18(8):1308-1331.

Li L, Liu R, Zhao J, Cai H, Yang Z. Effects of nb addition on glass-forming ability, thermal stability and mechanical properties of Ti-based bulk metallic glasses. *Rare Metal Materials and Engineering* (2014)43(8):1835-1838.

Li Y, Zhang W, Dong C, Qiang J, Xie G, Fujita K, Inoue A. Glass-forming ability and corrosion resistance of Zr-based Zr–Ni–Al bulk metallic glasses. *Journal of alloys and compounds* (2012) 536:S117-S121.

Liang W, Shen J, Sun J. Effect of si addition on the glass-forming ability of a nitizralcu alloy. *Journal of alloys and compounds* (2006) 420(1-2):94-97.

Liaw PK, Miller M. *Bulk metallic glasses: An overview.* New York, NY: Springer; 2007.

Lin CJ, Spaepen F. Fe-b glasses formed by picosecond pulsed laser quenching. *Applied Physics Letters* (1982)41(8):721-723.

Liu CT, Lu ZP. Effect of minor alloying additions on glass formation in bulk metallic glasses. *Intermetallics* (2005) 13(3-4):415-418.

Liu S, Su X, Pan P, Zhang L, Hu Y, Tan H, Li H. Neutrophil extracellular traps are indirectly triggered by lipopolysaccharide and contribute to acute lung injury. *Scientific reports* (2016) 6(1):1-8.

Löffler JF. Bulk metallic glasses. *Intermetallics* (2003) 11(6):529-540.

Louzguine Luzgin DV, Inoue A. Bulk metallic glasses: Formation, structure, properties, and applications. *Handbook of magnetic materials* (2013) 21:131-171.

Louzguine Luzgin DV, Louzguina Luzgina LV, Churyumov AY. Mechanical properties and deformation behavior of bulk metallic glasses. *Metals* (2012) 3(1):1-22.

Louzguine DV, Inoue A. Evaluation of the thermal stability of a $Cu_{60}Hf_{25}Ti_{15}$ metallic glass. *Applied physics letters* (2002) 81(14):2561-2562.

Lu H, Zhang L, Gebert A, Schultz L. Pitting corrosion of Cu-Zr metallic glasses in hydrochloric acid solutions. *Journal of alloys and compounds* (2008) 462(1-2):60-67.

Lu Z, Liu C. Role of minor alloying additions in formation of bulk metallic glasses: A review. *Journal of materials science* (2004) 39:3965-3974.

Manjunatha S, Reddy BC, Manjunatha H, Vidya Y, Sridhar K, Seenappa L, Gupta PD. Effect of calcination temperature on structural, antibacterial, radiation shielding and magnetic properties of cubic spinel cobalt nanoferrite. *Materials Science and Engineering: B* (2022) 286:115978.

Men H, Hu Z, Xu J. Bulk metallic glass formation in the Mg–Cu–Zn–Y system. *Scripta Materialia* (2002) 46(10):699-703.

Naka M, Hashimoto K, Masumoto T. High corrosion resistance of chromium-bearing amorphous iron alloys in neutral and acidic solutions containing chloride. *Corrosion* (1976) 32(4):146-152.

Nakai Y, Yoshioka Y. Stress corrosion and corrosion fatigue crack growth of Zr-based bulk metallic glass in aqueous solutions. *Metallurgical and Materials Transactions A* (2010) 41:1792-1798.

Nieh T, Wadsworth J. Homogeneous deformation of bulk metallic glasses. *Scripta Materialia* (2006) 54(3):387-392.

Nishiyama N, Amiya K, Inoue A. Novel applications of bulk metallic glass for industrial products. *Journal of Non-Crystalline Solids* (2007) 353(32-40):3615-3621.

Niu HY, Cao FF, Deng KK, Nie KB, Kang JW, Wang HW. Microstructure and corrosion behavior of the as-extruded Mg-4Zn-2Gd-0.5Ca alloy. *Acta Metallurgica Sinica (English Letters)* (2020) 33:362-374.

Nu NTN, Luong T. Potential applications of metallic glasses. *Int. J. Sci. Environ. Technol.* (2016) 5(4):2209-2216.

Pan X, Zhang H, Zhang Z, Stoica M, He G, Eckert J. Vickers hardness and compressive properties of bulk metallic glasses and nanostructure-dendrite composites. *Journal of materials research* (2005) 20(10):2632-2638.

Pang S, Zhang T, Asami K, Inoue A. Effects of chromium on the glass formation and corrosion behavior of bulk glassy Fe-Cr-Mo-cb alloys. *Materials Transactions* (2002) 43(8):2137-2142.

Park E, Chang H, Kim D, Ohkubo T, Hono K. Effect of the substitution of ag and ni for cu on the glass forming ability and plasticity of $Cu_{60}Zr_{30}Ti_{10}$ alloy. *Scripta Materialia* (2006) 54(9):1569-1573.

Park E, Kim D. Design of bulk metallic glasses with high glass forming ability and enhancement of plasticity in metallic glass matrix composites: A review. *Metals and Materials International* (2005) 11:19-27.

Park E, Kim D, Ohkubo T, Hono K. Enhancement of glass forming ability and plasticity by addition of nb in Cu–Ti–Zr–Ni–Si bulk metallic glasses. *Journal of non-crystalline solids* (2005) 351(14-15):1232-1238.

Park J, Wang G, Pauly S, Mattern N, Kim D, Eckert J. Ductile ti-based bulk metallic glasses with high specific strength. *Metallurgical and Materials Transactions A,* (2011) 42:1456-1462.

Patel SK, Swain BK, Behera A, Mohapatra SS. Metallic glasses: A revolution in material science. *Metallic Glasses* (2020).

Peker A, Johnson WL. A highly processable metallic glass: $Zr_{41.2}Ti_{13.8}Cu_{12.5}Ni_{10}Be_{22.5}$. *Applied Physics Letters* (1993) 63(17):2342-2344.

Peter W, Liaw P, Buchanan R, Liu C, Brooks C, Horton Jr J, Wright J. Fatigue behavior of $Zr_{52.5}Al_{10}Ti_5Cu_{17.9}Ni_{14.6}$ bulk metallic glass. *Intermetallics* (2002) 10(11-12):1125-1129.

Qin C, Zhang W, Asami K, Kimura H, Wang X, Inoue A. A novel cu-based BMG composite with high corrosion resistance and excellent mechanical properties. *Acta materialia* (2006) 54(14):3713-3719.

Qin FX, Zhou Y, Ji C, Dan ZH, Xie GQ, Yang S. Enhanced mechanical properties, corrosion behavior and bioactivity of ti-based bulk metallic glasses with minor addition elements. *Acta Metallurgica Sinica (English Letters)* (2016) 29:1011-1018.

Qin F, Yoshimura M, Wang X, Zhu S, Kawashima A, Asami K, Inoue A. Corrosion behavior of a Ti-based bulk metallic glass and its crystalline alloys. *Materials transactions* (2007) 48(7):1855-1858.

Qiu C, Liu L, Sun M, Zhang S. The effect of nb addition on mechanical properties, corrosion behavior, and metal-ion release of ZrAlCuNi bulk metallic glasses in artificial body fluid. *Journal of Biomedical Materials Research Part A: An Official Journal of The Society for Biomaterials, The Japanese Society for Biomaterials, and The Australian Society for Biomaterials and the Korean Society for Biomaterials* (2005) 75(4):950-956.

Reddy S, Bhattacharjee P, Murty B. The status of bulk metallic glass and high entropy alloys research. *Future Landscape of Structural Materials in India* (2022) 233-278.

Russo J, Romano F, Tanaka H. Glass-forming ability: Fundamental understanding leading to smart design. *Institute of Industrial Science, The University of Tokyo* (2018).

Samavatian M, Gholamipour R, Samavatian V, Farahani F. Effects of nb minor addition on atomic structure and glass forming ability of $Zr_{55}Cu_{30}Ni_5Al_{10}$ bulk metallic glass. *Materials Research Express* (2019) 6(6):065202.

Sarac B. *Microstructure-property optimization in metallic glasses*. New York City, US: Springer; 2015.

Schuh CA, Hufnagel TC, Ramamurty U. Mechanical behavior of amorphous alloys. *Acta Materialia* (2007) 55(12):4067-4109.

Scully JR, Gebert A, Payer JH. Corrosion and related mechanical properties of bulk metallic glasses. *Journal of Materials research* (2007) 22(2):302-313.

Shao L, Zhang T, Li L, Zhao Y, Huang J, Liaw PK, Zhang Y. A low-cost lightweight entropic alloy with high strength. *Journal of Materials Engineering and Performance* (2018) 27:6648-6656.

Stolle A, Ranu B. *Ball milling towards green synthesis: Applications, projects, challenges*. Cambridge, UK: Royal Society of Chemistry; 2015.

Suñol J, Clavaguera N, Clavaguera Mora M. Comparison of Fe–Ni–P–Si alloys prepared by ball milling. *Journal of non-crystalline solids* (2001) 287(1-3):114-119.

Suryanarayana C, Inoue A. Thermal stability of metallic glasses. *Bulk Metallic Glasses* (2011) 208-211.

Suryanarayana C, Inoue A. Iron-based bulk metallic glasses. *International Materials Reviews* (2013) 58(3):131-166.

Suryanarayana C, Inoue A. *Bulk metallic glasses*. Boca Raton: CRC press; 2017.

Tan C, Jiang W, Zhang Z, Wu X, Lin J. The effect of Ti-addition on the corrosion behavior of the partially crystallized Ni-based bulk metallic glasses. *Materials Chemistry and Physics* (2008) 108(1):29-32.

Tsai PH, Lee CI, Song SM, Liao YC, Li TH, Jang JSC, Chu JP. Improved mechanical properties and corrosion resistance of mg-based bulk metallic glass composite by coating with Zr-based metallic glass thin film. *Coatings* (2020) 10(12):1212.

Wang D, Li Y, Sun B, Sui M, Lu K, Ma E. Bulk metallic glass formation in the binary cu–zr system. *Applied Physics Letters* (2004) 84(20):4029-4031.

Wang WH, Dong C, Shek C. Bulk metallic glasses. *Materials Science and Engineering: R: Reports,* (2004) 44(2-3):45-89.

Wang YL, Xu J. Ti (Zr)-Cu-Ni bulk metallic glasses with optimal glass-forming ability and their compressive properties. *Metallurgical and Materials Transactions A,* (2008) 39:2990-2997.

Yang Y, Cheng B, Lv J, Li B, Ma M, Zhang X, Liu R. Effect of ag substitution for ti on glass-forming ability, thermal stability and mechanical properties of Zr-based bulk metallic glasses. *Materials Science and Engineering: A,* (2019) 746:229-238.

Yavari AR, Lewandowski J, Eckert J. Mechanical properties of bulk metallic glasses. *MRS Bulletin* (2007) 32(8):635-638.

Zhang C, Li Q, Xie L, Zhang G, Mu B, Chang C, Ma X. Development of novel Fe-based bulk metallic glasses with excellent wear and corrosion resistance by adjusting the Cr and Mo contents. *Intermetallics,* (2023) 153:107801.

Zhang M, Song Y, Lin H, Li Z, Li W. A brief introduction on the development of Ti-based metallic glasses. *Frontiers in Materials* (2022) 8:575.

Zhang T, Inoue A. Thermal stability and mechanical properties of bulk glassy Cu-Zr-Ti-(Nb, Ta) alloys. *Materials transactions* (2002) 43(6):1367-1370.

Zhang X, Ma E, Xu J. Mechanically alloyed multi-component mo-based amorphous alloys with wide supercooled liquid region. *Journal of non-crystalline solids* (2006) 352(38-39):3985-3994.

Zhang Y, Yao J, Zhao X, Ma L. Ti substituted Ni-free $Zr_{65-x}Ti_xCu_{17.5}Fe_{10}Al_{7.5}$ bulk metallic glasses with significantly enhanced glass-forming ability and mechanical properties. *Journal of Alloys and Compounds* (2019) 773:713-718.

Zhang Z, Eckert J, Schultz L. Difference in compressive and tensile fracture mechanisms of $Zr_{59}Cu_{20}Al_{10}Ni_8Ti_3$bulkmetallic glass. *Acta Materialia* (2003) 51(4):1167-1179.

Chapter 4

Synthesis and Properties of Nanostructured Metallic Glass

Shiv Prakash Singh*
Center for Advanced Ceramics Materials,
International Advanced Research Institute of Powder Metallurgy and New Materials (ARCI), Hyderabad, India

Abstract

Glass is a thermodynamically metastable and structurally disordered material. Metallic glass has attracted special attention for its interesting mechanical, thermal, electrical, and magnetic properties. Conventional bulk metallic glasses (CBMG) consist of two or more elements and are prepared by the conventional melt-spin technique. CBMG does not show any microstructure and is homogeneous. Recently, nanostructured metallic glass (NMG) attracted special attention for its unusual structure and other fascinating mechanical, thermal, magnetic, and electrical properties. The NMG is primarily prepared by electrochemical, chemical route synthesis, and physical vapor deposition techniques such as inert gas condensation, sputtering, pulsed laser deposition, etc. The atomic structure of the NMG is different than the CBMG. The NMG consists of two features, a glassy core and glass-glass interfaces. Due to this unique structural feature of NMG, it shows exciting properties compare to its CBMG. The current understanding of the NMG is at the preliminary stage, and it needs more studies to explore its fascinating properties. It is believed that nanostructured metallic glasses bring a next-generation revolution in nanotechnology.

* Corresponding Author's Email: spsingh67@gmail.com.

In: Properties and Uses of Metallic Glass
Editor: Shiv Prakash Singh
ISBN: 979-8-89113-569-7

Keywords: metallic glass, nanostructured, synthesis, properties

Introduction

We have found metals and alloys around us in the crystalline form. However, in the 1960s, there is a discovery of a new type of material known as metallic glass (MG) that open a new avenue for the materials science community. Metallic glasses are amorphous and thermodynamically metastable materials with interesting properties compared to counter crystalline materials. The high strength and corrosion resistance nature of the MG makes it attractive for many applications in the area of biomedical, space crafts, sports equipment, etc. (Halim et al., 2021). There are much progress has been made in the research and development of the MG in the last decades. The first MG was produced with two elements, and now it has been manufactured in multi-component MG compositions by adopting different synthesis processes (Li et al., 2019; Halim et al., 2021; Jiang et al., 2022). The most important criterion for the formation of the MG is its glass-forming ability (GFA) which hindered the crystallization tendency and resulted in the uniform amorphous material. A good MG can be obtained with high GFA, fast cooling rate, suitable glass forming compositions, and optimized glass processing parameters. Synthesis of the bulk metallic glass (BMG) is required higher GFA alloys that can be cast completely in an amorphous phase, and the cast dimension is at least 1 mm (Schroers, 2010). Technologically more relevant magnesium, lanthanum, and zirconium-based BMG was discovered by the Inoue research group from the Tohoku University, Japan, during the 1990s (Inoue et al., 1988, 1989, 1990, 1993). In the later phase, BMG is produced using a wide range of elements in the glass compositions and using different preparation methods.

There is always demand for new materials that can be produced either by adjustment in the composition or optimized processing parameters. At present, nanostructured materials have evolved as potential materials for several scientific and technological applications. It has been a growing interest to explore the new properties of nanostructured materials using advanced characterization techniques. Nanostructured metallic glass (NMG) is one of the potential materials with fascinating properties for many applications. There are few reports on the nanoglass, a new kind of NMG that comprises two types of structural features, such as glassy core and glass-glass interface (Gleiter, 2008, 2016; Ivanisenko et al., 2018). Nanoglass is made by compaction of glassy nanoparticles and creates a glass-glass interface between them. It is found that the atomic arrangements at the glass-glass interfaces are different than the glass core. The nanoglass structure is comparable with the nanocrystalline materials that have crystalline grains and grain boundaries. The atomic arrangements of the grain and grain boundaries are different and the properties can be tuned by changing the grain boundary densities. A schematic presentation of the difference in the physical properties of the crystal, bulk magnetic glass, and nanostructured metallic glass is displayed in Figure 1.

There are different synthesis processes, such as chemical route synthesis, physical vapor deposition, electrochemical, etc., for the preparation of the nanostructured materials that will be discussed in the subsequent sections. The synthesis processes are selected based on the chemical compositions and targeted nanostructured features. The NMG is characterized using advanced characterization tools. It needs a meticulous analysis of the experimental results to state the structure and properties of the NMG. Moreover, the different properties, such as structural, thermal, mechanical, magnetic, catalytic, etc., of the NMG are discussed here in detail. In this review, it is highlighted that there is limited applications are reported for NMG, but it is considered as a potential material for many prospective applications in the future.

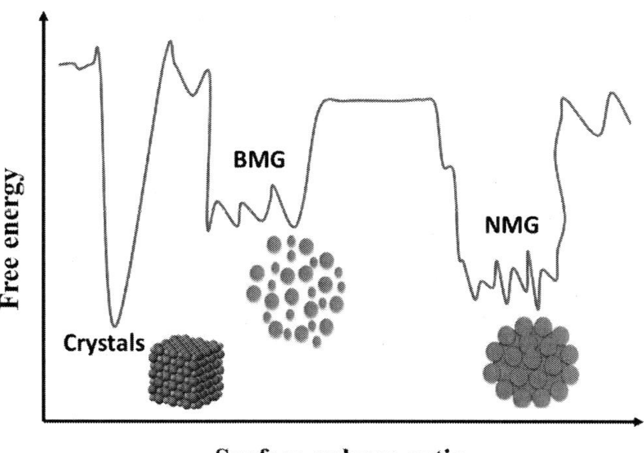

Figure 1. Schematic presentation of physical properties differences of the crystal, bulk metallic glass (BMG), and nanostructured metallic glass (NMG).

Synthesis

Nanostructured metallic glass can be synthesized either in a top-down approach or a bottom-up approach. However, a top-down approach is the most common method that developing nanostructures in bulk metallic glasses. The bottom-up approach is not used popularly and hence is less discussed in the literature. Metallic glasses produced by the bottom-up approach are mainly involved in the chemical route synthesis process. Here we have discussed the different methods for the synthesis of the nanostructured metallic glass as follows. The first two sections discussed the synthesis of BMG and further processed to produce NMG. Other sections are presented the single-step process to produce NMG.

Melt Spinning Technique

Melt spun is the conventional method for the preparation of metallic glass, particularly in the ribbon form. A binary to multi-components including Zr, Cu, Ni, Si, Pd, etc. in the glass compositions that shows good glass-forming ability is used in the melt-spinning synthesis process. In this method, a metallic alloy ingot is prepared of the desired composition using the arc-melting method with good homogeneity throughout the sample. The ingot is further melted at high temperature to a liquid state and drained through the nozzle leading to a rapid quench on the metallic roller that continuously rolls at a certain speed. Here high temperature melts cooled down to room temperature through fast heat dissipation from glass melt to the metallic roller of a bigger surface area under air cooling. The ribbon is come out from the roller surface by centrifugal force. A schematic presentation is displayed in Figure 2. Here a cooling rate of 10^6 K/second is achieved during the quenching process (Budhani et al., 1982). However, a high quenching rate can be achieved by tuning the different parameters such as the ejection pressure, nozzle width, distance between the nozzle and roller, and roller speed. The width and thickness of the ribbon were obtained in the range of 2-3 mm and ~ 30 μm respectively. The prepared metallic glass ribbon undergoes further treatments to develop nanostructures in the glass matrix. Mechanical deformations such as bending, rolling, or shearing under certain loads are one of the known techniques for the development of nanostructured glass. In this case, the nanostructure can be varied with the change in the various process parameters such as applied load, time for deformation, number of cycles, etc. Another method for the nanostructure features in the metallic glass ribbon is the thermal treatment at different temperatures and time duration. Due to optimized thermal treatment, the atomic arrangements are changed that leads to interesting nanostructures in the ribbon. The thermal treatment can be provided by conventional resistive heating or by lasers of different wavelengths.

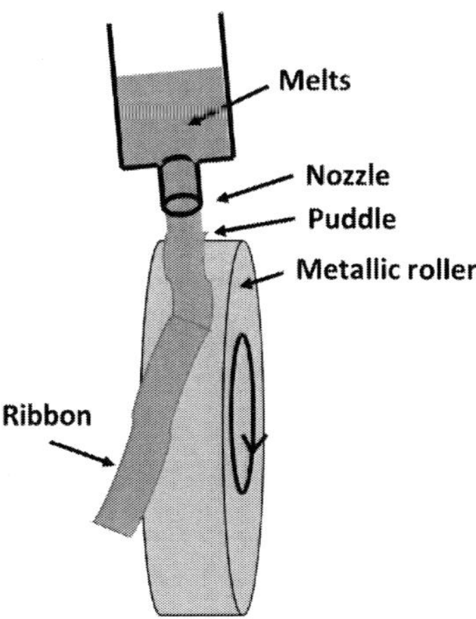

Figure 2. Schematic presentation of the preparation of metallic glass ribbon from the metallic alloy melts on the rotating roller.

Melt Splat Quenching Process

In 1959, the first metallic glass preparation of gold-silicon alloys was reported by Klement et al., from the California Institute of Technology using the melt splat quenching method (Klement et al., 1960). This method achieves a high cooling rate for the metallic glass preparation. In this process, the glass composition consists of two or more elements. A pre-synthesized alloy ingot is subjected to a high melting temperature to obtain the liquid melt. The glass melt is poured in droplet form onto a massive copper substrate that continuously cools with water circulation and is instantly quenched by transferring the heat from the glass melt to the copper roller. The melt drop spread and flattened into a sheet of a few micrometer thicknesses on the copper substrate. The sheet has very good surface contact with the copper roller that helps in quick heat transfer and quenching process. In this process, the cooling rate during quenching is much faster than in the conventional solidification process. The parameters that decide the nature of the final product are the size and viscosity of the melt drop, the distance between the melt drop and the substrate, the thickness of the impacted sheet, etc. The glassy nature of the final sample depends on the above parameters that control the atomic arrangements and microstructure of the glass. Nanostructured metallic glass can be produced from the splat quench glass by changing the different parameters during the process or through further thermal or mechanical treatments after the preparation of the metallic glass sheet.

Physical Vapor Deposition (PVD)

The PVD method is widely used for the fabrication of the uniform thin film of nanostructured metallic glass. This method is very flexible to choose a wide range of desired compositions. In this method, a thin layer of the metallic glass composition is deposited on the substrate under a high vacuum. In this method, atoms are removed from the target materials using high-energy ions or thermal energy under a vacuum. The evaporated atoms from the surface of the target are moved toward the substrate and deposited on it. The atomic structure of the deposited films can be tuned with the evaporation rate, inert gas flow rate, substrate temperature, etc. Several PVD techniques are used such as magnetron sputtering (either DC or RF mode), inert gas condensation (IGC), electron beam epitaxy, and laser ablation based on the energy source for the high temperature. In most PVD methods, atoms are in the vapor state and quenched on the substrate surface, resulting in thin films. Magnetron sputtering is the most common PVD technique used to produce a thin film of metallic glass. The thin films are obtained with high purity, good adhesion, and high uniformity. These are two types of magnetron sputtering based on the working principles such as (i) direct current (DC) magnetron and (ii) radio frequency (RF) magnetron sputtering systems. RF magnetron has an advantage over the DC magnetron because it can be applied to electrically poor conductive target materials, and hence, in principle, any target materials can be sputter with RF magnetron sputtering systems. In both processes, a collision between incident particles/ions and targets takes place. In general, high-speed sputtering is executed at a low pressure that stimulates the increase in the ionization rate (plasma formation) of the gas. The incident particle collides with the target atom and experiences a complicated scattering process in the target. The part of scattering transmitted the momentum to the target atom, and the other part of scattering directed certain atoms near the target surface that gained sufficient momentum for outward motion and resulted in sputtering out of the target. In the case of magnetron sputtering, it increases the plasma density on the target surface by introducing a magnetic field that constrains the charged particles within the magnetic field to accelerate the sputtering rate. A schematic presentation for the preparation of thin films using magnetron sputtering is displayed in Figure 3. In the IGC method, nanopowders of 5-20 nm size are synthesized by colliding with the helium atoms that flow into the vacuum chamber (Gleiter, 2008, 2016). The nanosized powders are deposited on the cold finger that is continuously cooled by the nitrogen at the center of the vacuum chamber. The powders are compacted in-situ in the form of pellets without breaking the vacuum. The IGC process can be performed either by thermal evaporation or sputtering. In the IGC process, a cooling rate can be attained 10^{12} °C/minute. In this method, a non-conventional glass composition can be prepared. The PVD method is the most popular method for preparing the nanostructured metallic glasses. In recent years, several studies have been published on the synthesis of metallic nanoglass using PVD techniques (Bag et al., 2020; Katnagallu et al., 2020; Singh et al., 2020b; Sharma et al., 2021).

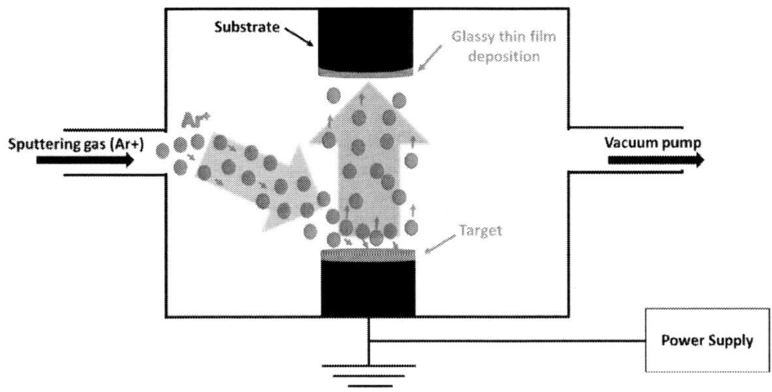

Figure 3. Schematic presentation of the preparation of uniform metallic glass thin film by physical vapor deposition (PVD) method in the sputtering vacuum chamber.

Electro-Synthesis

The preparation of high-purity nano-crystalline materials has been reported by an electrodeposition process that produced a large volume fraction of grain boundaries (i.e., crystal/crystal interfaces) in the materials (Natter et al., 1998; Lu et al., 2000). There are few reports available on the preparation of metallic glass through the electro-synthesis route. In this case, an electric potential is applied to the anode and cathode end in the presence of an electrolyte, and desired metal salts are diluted in the solution either in an aqueous or alcoholic medium. This type of metallic glass is prepared at room temperature and is very cost-effective. However, there is very few compositions are reported that can be formed glass in this method. Guo et al., have prepared Ni-P nanoglass by specially designed multi-phase pulse electrodeposition method (Guo et al., 2017). Here for the Ni-P nanoglass synthesis, a conventional three-electrode cell and an electrochemical workstation were used for the electrodeposition. The final composition of $Ni_{78}P_{28}$ nanoglass was obtained by the electrochemical reaction of among the electrolytes $NiSO_4.6H_2O$, $NiCl_2.6H_2O$, and H_3PO_3. In this process, H_3BO_3 is used as a supporting electrolyte and a p^H buffer solution. The whole synthesis process was operated in three steps of applied voltage, (i) 9V, (ii) 5V, and (iii) 0V. The thin film of the Ni-P nanoglass was deposited on the cleaned copper substrate.

High Pressure Torsion (HPT)

HPT method is a mechanical method that forms nanostructured metallic glass through plastic deformation in bulk metallic glass materials (Zhilyaev and Langdon, 2008). In the HPT process, bulk metallic glass is subjected to high pressure (about 5-10 GPs) and then mechanical torsion is applied. During this process, there is an increase in the local temperature up to 1000 ^{o}C or even more. This type of deformation results in the mobility of the atoms from their equilibrium state to the non-equilibrium state and forms nanostructured glass. This plastic deformation is in the form of shear bands. The atomic arrangements in the shear bands are different than the un-deformed areas. Such deformations exhibit interesting properties, such as mechanical, magnetic, thermal, and structural. This method is not suitable for the bulk production of nanostructured metallic glass and is limited to the lab scale. Wang et al., have demonstrated the plastic deformation induced by HPT in Au-based BMG that generated a high density of shear bands (Wang et al., 2011). These shear bands have reduced ordered atomic structures and high free volume. In another work, Wang et al., reported the synthesis of BMG of $Zr_{50.7}Cu_{28}Ni_9Al_{12.3}$ that was further subjected to the HPT to improve the ductility of the material (Wang et al., 2012).

Extrusion

Extrusion is another type of mechanical method to produce nanostructured metallic glass. In this case, high pressure is applied through the channels at high shear that bring the deformation within the material at the atomic level. In this method, a large quantity of nanostructured metallic glass can be produced. Equal channel angular pressing (ECAP) is a well-known process to produce nanostructured materials following the extrusion principle (Ebrahimi et al., 2015; Ebrahimi and Gode, 2017). Bulk samples in the rod- and plate shape are pushed through ECAP that are angularly extruded and placed in the angular hole. Here, the flow is constrained in all directions. However, under externally applied pressure, the flow is permitted in one direction. The nature of plastic deformation depends on several factors of the ECAP process, such as pressure, channel size and curvature angle, pressing speed, temperature, material composition, etc. (Valiev and Langdon, 2006). The properties of the NMG final products processed by ECAP can be altered by varying the process parameters and is an interesting area of research. Tong et al., have introduced residual elastic strain in the $(La_{0.5}Ce_{0.5})_{65}Co_{25}Al_{10}$ and $Zr_{52.5}Cu_{17.9}Ni_{14.6}Al_{10}Ti_5$ bulk metallic glass using the ECAP method (Tong et al., 2013). It is found that the origin of the residual elastic strain is due to the formation of shear bands and deformation occurred by the ECAP process. They performed the ECAP in a tungsten-carbide die with a channel angle of 160°. The whole process was carried out in the gasoline medium under 1.2 GPa hydrostatic pressure and the applied nominal normal stress was 14% lower than the elastic limit.

Characterization

NMG shows interesting properties compared to BMG of the same chemical composition, and hence, it attracts the attention of many researchers. To explore its different properties and structure, a detailed experimental analysis is required. There are several characterization tools are available to characterize the NMG and some of them are discussed below.

Structural Characterization

It is known that glass is an atomically disordered amorphous material and does not show defined atomic structural features. X-ray diffraction (XRD) pattern is a primary tool to characterize the nature of the glass. The XRD experiments can be performed either using bulk or powder forms. If the sample is reactive with the air, then the sample can be sealed in a special holder of a glass dome inside the glove box. If the XRD experiment is performed using the sealed sample holder, then it will be good to do the XRD of the sample holder without the sample before the real experiment to confirm any diffraction signal from the sample holder. In this technique, one can confirm the material is either amorphous or crystalline. Depending on the need for structural information and chemical compositions of the metallic glass, different types of X-ray energy sources are used, such as Cu, Ag, Mo, and Ga to confirm the amorphous nature. The cu source is the most common source for X-rays of the wavelength of 1.54 Å is used in the XRD experiment. However, certain compositions, such as Fe-containing metallic glass samples do not perform the Cu XRD as the Cu k-alpha energy is similar to the Fe k-alpha energy and has a fluorescence effect on the sample. Thus, there is a possibility that obtained diffraction pattern may not provide precise information about the sample. High energy X-ray provides more accurate and high-definition information on the atomic arrangement in the metallic system. For a better understanding of the atomic arrangements and the pair distribution, researchers used X-rays in synchrotron facilities. These X-rays are comparatively higher in energy. In addition to the structural information, we can also analyze the structural relaxation and excess free volume in the glassy phase of the BMG (Mao et al., 1995; Yavari et al., 2005). In some special cases, researchers also use neutron diffraction to elucidate the metallic glass structure using a synchrotron facility.

Some other structural characterizations are available for direct observation of the structural features such as electron microscopy and atom probe tomography (APT). Scanning electron microscope (SEM) and transmission electron microscope (TEM) are the most popular electron microscope techniques that provide information on the homogeneity of the microstructure throughout the sample. SEM provides surface information that shows no crystallinity in the specimen at the nanoscale. Using the electron diffraction X-ray (EDX) attachment to the SEM provides the chemical composition of the sample and also helps in the mapping of the chemical gradients. For the SEM experiment, a bulk sample of a few microns to mm in size is used at diffraction magnifications.

TEM is a very precise electron microscopy tool that provides information at the atomic level. TEM study of the glass shows a homogeneous feature without any crystallinity in the specimen. For the TEM experiment, either bulk or powder samples can be used. However, for the bulk sample, additional sample preparation is required. For the bulk sample, a specimen has to be milled to make it very thin, up to 100 nm using a focused ion beam (FIB) or by traditional electron milling. An electron beam passed through different optics components to focus on the sample and passed through it that provides information about the sample at the atomic level. There are other attachments are attached with the TEM for the precise analysis of the structure and chemical compositions of the sample. EDX is attached to analyze the chemical composition. There is a separate attachment for the selected area electron diffraction (SAED) experiment at the nanoscale to examine the nature of the sample whether it is amorphous or crystalline. The metallic glass is amorphous and hence it shows a diffused ring, whereas the presence of crystalline dots for the crystalline sample in the SAED image. Recently, researchers have analyzed the radial distribution function (RDF) for the atomic arrangement using the SAED data. To have a clear understanding of the surface of the TEM sample, a high-angle annular dark field scanning transmission electron microscopy (HAADF-STEM) is carried out to see any structural features on the surface of the sample. HAADF-STEM is a STEM technique that uses inelastically scattered electrons at high angles with an annular dark-field detector.

To understand the 3D atomic structure of metallic glass, the most advanced technique is atom probe tomography (APT). APT provides information about the atomic configuration and chemical composition at the nanoscale. The very tiny sample of a few nanometers (~100-200 nm) is prepared for the APT experiment by a lift-out method using the FIB instrument. Annular milling was used to obtain needle-shaped APT samples with a typical radius of curvature of about 100 nm. The Ga^+ ion is used for the ion milling of the sample to nanometer thin. The final step of the FIB "clean-up" procedure is performed at low voltage (~5 kV) to ensure the least content of uncontrolled impurities (Ga impurities, if any). The preparation of the sample needs special caution to select the parameters that assure the original structure of the sample during preparation. There are different methods to carry out the APT measurements such as laser pulsing mode (typical wavelength 355 nm, pulse frequency ~100 kHz, sample temperature ~60 K, pulse energy ~50 pJ, evaporation rate about 0.5%), and field-assisted evaporation by applying high potential. The APT data is used to reconstruct the atoms with the software package. Finally, it presents the 3D arrangements of the atoms of the sample. This 3D data analyzes in many ways to understand the chemical segregation, deformations such as shear bands, and diffusion mechanism of different elements present in the sample.

Mechanical Properties Evaluation

Metallic glasses show interesting mechanical properties compared to crystalline materials. It shows ductility, good strength, and high corrosion resistivity. Mechanical properties such as hardness, strength, and fracture toughness are the common properties measured using indenter, three/four-point bending mechanical measurement instruments (Czichos et al., 2006). IIowever, for the nanostructured metallic glass, more sophisticated characterization tools are applied such as nano-indentation, elastic moduli, tensile / compression testing using bulk samples and/or micro-pillars fabricated by focused ion beam (FIB) instrument (Phaneuf, 1999; Giannuzzi and Stevie, 2005; Volkert and Minor, 2007; Gierak, 2009; Bassim et al., 2014). The pillars are milled by a top-down preparation method using a Ga^+ ion beam to micron size in diameter. FIB provides a precise dimension for the measurement of deformation at the atomic level during mechanical testing (Phaneuf, 1999; Volkert and Minor, 2007). For example, the stress-strain curve plotting for the compressing testing by applying load on the micro-pillars of the nanostructured metallic glass that provides the yield strength of the materials (Tabbakh et al., 2021; Basak et al., 2023). Another type of mechanical property, elastic moduli measured to know the stiffness of the materials (W. Arnold, R. Birringer, C. Braun, H. Gleiter, H. Hahn, S. H. Nandam, 2020). The conventional method for the measurement of elastic moduli is based on ultrasound velocities. The ultrasound velocities are determined by two types of methods. The first one is the pulse-echo method using a single delay-line transducer or the ultrasonic transmission method to measure the time delay through a sample placed between two delay-line transducers. This procedure is also termed as wide-band excitation. The second method used the rf-carrier system as an electronic transmitter whose carrier frequency oscillates from 20 to 200 MHz that is adjusted to the resonance frequency of the transducer. In this case, the number of rf-oscillations is typically 3-5. This second technique is known as narrow-band excitation. The wide-band excitation is used for the samples which are sufficiently thick. In this case, pulse-echo tests with well-separated echo patterns are permitting one to measure the time-of-flight of the signals. Elastic moduli and Poisson ratios can be calculated by knowing the specimen's densities. The nanostructured metallic glass samples are thick enough (~120 μm) to get the echo patterns with clearly distinguished separate echoes (W. Arnold, R. Birringer, C. Braun, H. Gleiter, H. Hahn, S. H. Nandam, 2020). Nano-indentation experiments are carried out using a diamond Berkovich tip attached on a nanoindenter head. The tip area function was standardized on fused silica before the experiment. The measurements were accomplished at room temperature under load-control mode, with a continuous stiffness measurement method that gives the values of hardness and elastic modulus with increasing penetration depth. During the measurement, maximum indentation depth was restricted to 500 nm. According to the Oliver and Pharr method, the values of hardness and elastic modulus are calculated for depths below 10% of sample thickness (i.e., ~ 50-100 nm) to avoid substrate influence (Oliver and Pharr, 2004).

Thermal Properties Characterization

The metallic glass is a thermodynamically metastable material, and hence, it is important for its thermal properties analysis. The metallic glass shows glass transition temperature which is characteristic of the glassy nature of the materials. Other thermal properties such as crystallization and melting temperatures, specific heat, and thermal conductivity. These thermal properties are demonstrated through thermal transport via phonon-phonon interaction (energy transfer). Glass forming ability is an important factor to understand the stability of the glass structure under thermal energy application. Differential scanning calorimetry (DSC) is the most popular instrument to measure the glass transition, crystallization temperatures, and melting temperature at different heating rates. In DSC measurement, a few mg (10-20 mg) weight of sample either in powder or bulk form is used. In this case, two sample holders are present that contain two crucibles with a lid. These small crucibles are either alumina or platinum based on the nature of the sample and the objective of the experiment. To understand the whole thermal history of the sample, a sample is heated till its melting temperature. The distinct exothermic peak is considered as a crystallization temperature and the endothermic peak indicates the melting temperature. Glass transition temperature is defined as an onset temperature range of the first exothermic peak. This exothermic peak is appearing as a weak and broad peak. The glass transition temperature is determined by drawing two slopes from the two curvatures of the peak. The meeting point of these two slops is considered a glass transition temperature. In some glass compositions, the glass transition temperature is not well defined. In such a situation, a higher heating rate can be helpful to see the peak in the DSC thermogram. Glass transition temperature is an important parameter to define an amorphous material as glass. The temperature difference (ΔT) between glass transition and crystallization temperatures is denoted as a glass-forming ability (GFA) of the glass. The higher ΔT gives higher GFA and it is considered as a higher thermal stability glass. The glass relaxation indicates how fast glass relaxes from its thermodynamic non-equilibrium state to the equilibrium state. This can be evaluated with the help of the specific heat of the metallic glass that is measured by DSC.

Magnetic and Electrical Properties Characterization

Some metallic glasses, like Fe, and Co-based show magnetic properties and are defined as soft or hard magnets depends on their magnetic strength. The magnetic properties such as magnetic moment, Curie temperature, and temperature-dependent magnetization are measured precisely by using a superconducting quantum interference device (SQUID) at different temperature ranges from cryogenic to high temperatures and at different external magnetic fields, typically 0-9 T. However, physical properties measurement system (PPMS) is also used conventionally for the measurement of magnetic and electrical properties. The internal magnetic hyperfine field splitting in most transition metal-based alloys relative to the magnetic moment of Fe that revealed the magnetic behavior associated with the structure and chemical environment of the magnetic species like Fe. The most common method is used to establish the structure-magnetic property relationship by nuclear methods such as Mössbauer spectroscopy. In this method, a solid sample is exposed to a gamma radiation beam, and a detector measures the beam intensity that is transmitted through the sample. The atoms resulting in the source for emitting the gamma rays need to be of the same isotope as the atoms in the absorbing sample.

Properties

The discovery of metallic glass has brought tremendous interest from researchers for its interesting properties compared to crystalline materials. Furthermore, nanostructured metallic glasses are a new kind of nanomaterials that need special attention to explore their different properties. The most important properties of the metallic glasses are provided below.

Structural

Metallic glass is defined as atomically disordered materials that show a lack of long-range atomic order like in crystalline materials. However, recent studies show medium-range atomic order for the nanostructure metallic glass like in the nanoglass. In the case of the metallic nanoglass, there are two glass structures, a glassy core and a glass-glass interface (Ivanisenko et al., 2018; Wang et al., 2019). It is found that the atomic arrangement of the glassy core is different than the glass-glass interface. There is density fluctuation at the glassy core and glass-glass interfaces. The interfaces revealed a higher free volume compare to the glassy core. This type of structural difference at the nanoscale in the nanoglass structure is reported by researchers. Ghafari et al., (2012b) have presented the structural analysis of $Fe_{90}Sc_{10}$ nanoglasses and rapidly quenched metallic glasses of the same chemical composition with high-resolution diffraction experiments using high-energy x-rays from a synchrotron radiation source. They found the short and intermediate-range atomic orders in a nanoglass which are different from the conventional metallic glass produced by the rapid quenching method from the melt. The atomic intermediate distance range for the nanoglass is closer packed than the conventional glass. The average atomic distance and the most plausible distance of nearest neighbors are a little different and depend on the compaction pressure. It is found that for the nanoglass, the first-nearest neighbor coordination number is decreased. This structural variation causes a change in the physical properties of the nanoglass compared to the conventional rapidly quenched glasses. Moreover, it is interesting to observe the stronger correlation between atoms at distances above 10 Å in the nanoglass than for the conventional glass. Thus, it can be understood that the interatomic distances and the coordination number can be manipulating to tuned the physical properties of the nanoglasses. Authors have presented the radial distribution function (RDF) of conventional metallic glass prepared by the rapidly quenched melt and nanoglass synthesized using inert gas condensation method of same composition of $Fe_{90}Sc_{10}$ alloys. They have shown that the nanoglass has additional structural features compared to the conventional metallic glass. It revealed that the nanoglass has a medium-range order atomic structure in addition to the short-range order similar to the conventional metallic glass. These additional structural features that arise in the nanoglass are predicted due to the glass-glass interfaces which have different atomic order compared to the glass core.

Mechanical

There is growing interest in metallic glass for structural application. Hence, substantial research is going on to improve the mechanical properties of the glass. It has been understood that nanostructured metallic glass has a different atomic structure compared to conventional metallic glass. Hence it is assumed that the mechanical properties of the nanostructured and conventional metallic glass will be different. In a nano-indentation study, Nandam et al. (2017) have found interesting results for the melt spun ribbon, nanoglass, and annealed nanoglass at 350°C/3h samples of the same $Cu_{50}Zr_{50}$ composition.

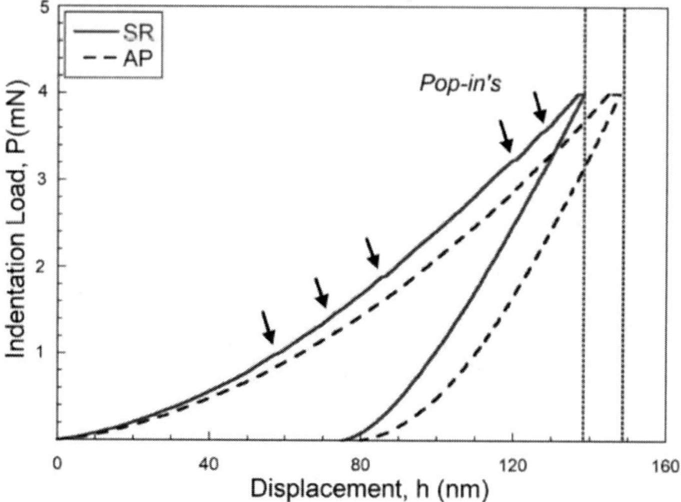

Figure 4. Illustrative micro-indentation curves of indentation load verses penetration depth (displacement) for the structurally relaxed (SR) and as-prepared (AP) Cu-Zr based nanoglass. The SR sample was prepared by using AP sample subjected to the annealing at 400 °C for 90 minutes. There is no pop-ins were found in the AP nanoglass but they were presented in SR sample that indicated by arrows. The image is reproduced from the reference (Sharma et al., 2021).

The as-prepared nanoglass and annealed nanoglass samples are shown smooth curves without any features of pop-ins that indicate the homogeneous deformation under the load during the indentation experiment. However, the melt-spun ribbon of the same composition shows distinct pop-ins, which revealed the presence of shear bands due to the inhomogeneous deformation. Another similar study on deformation behavior was carried out based on the micro-indentation of the as-prepared (AP) and structurally relaxed (SR) Cu-

Zr nanoglass (Sharma et al., 2021). The Cu-Zr nanoglass structurally relaxed by annealing the as-prepared sample at 400 °C for 90 minutes without changing their amorphous structure. Representative load versus penetration depth curves generated by micro-indentation for the AS and SR samples are depicted in Figure 4. The SR sample shows distinct displacement bursts (pop-ins). In contrast, the AP sample does not show such features. This observation revealed the localized change in the amorphous structure during the annealing process in the SR sample. It concluded that two different amorphous structures (AP and SR) show two different deformation phenomena due to the structural change.

The nanoglass structure has high free volume at the glass-glass interfaces compared to the melt-spun ribbons with homogeneous microstructure. The areas with high free volume act as nucleating sites for shear transformation zones (STZ) in the metallic glasses (Singh et al., 2020a). There is a large volume of interfaces present in the nanoglass, which transforms the STZs into the matured shear bands. Hence, deformation in the nanoglass samples shows pop-ins when subjected to the loads. Moreover, in another study, Singh et al., (2020a) reported the fast diffusion of the Tb in the shear band generated in the $Tb_{75}Fe_{25}$ nanoglass under severe plastic deformation using high-pressure torsion (HPT). There is a change in the microstructure and chemical compositions observed in the $Tb_{75}Fe_{25}$ nanoglass subjected to the HPT compared to the undeformed nanoglass. The undeformed $Tb_{75}Fe_{25}$ nanoglass shows a homogeneous microstructure in the TEM studies and there is no significant chemical composition fluctuation found in the APT results.

Thermal

The glass-forming ability during the cooling of a melt liquid determines the formation of crystalline and glassy phases. If the crystal nucleation and growth are diminished, the glass-forming ability is enhanced and results in glass formation. The glass transition temperature (T_g) and the crystallization temperature (T_c) are determined factors for the various glass applications. The thermal properties of the glass can be changed by modification in the chemical composition and the process parameters. Metallic glasses have a simple atomic packing structure and hence are considered a model system to study the thermal kinetics of glasses. Recent studies have shown that the new kind of nanostructured glass, termed nanoglass has different atomic arrangements at the core and the interfaces. It will be very interesting to explore the effect of the nanostructure on thermal behavior by controlling the glass-glass interfaces in the nanoglass (Wang et al., 2014).

Wang et al., (2014) have synthesized Au-based nanoglass that revealed kinetic and thermodynamic ultra-stability at low heating rates compared to the melt-spun ribbon counterpart of the same chemical composition. It is reported that the nanoglass demonstrates higher T_g of 20 K, higher T_c of 32 K, superior crystallization activation energy, slow crystal growth rate, and low enthalpy compared to the melt-spun ribbon. The relaxation during vapor phase deposition and the nanostructure two main factors that are contributed to the ultra-stability of the nanoglass. Nandam et al., (2017) have synthesized $Cu_{50}Zr_{50}$ metallic nanoglasses using sputtering in the inert gas condensation chamber. The study shows that the nanoglass comprises Zr-rich dense cores and Cu-rich interfaces with enhanced free volume. These structural features offer the low enthalpy and high T_g of the nanoglass compared to melt-spun ribbons of the same composition. Mohri et al., (2018) have produced thin films of the Ti-Zr-Cu-Pd system with a nanoglass microstructure using DC magnetron at different sputtering settings. The effect of sputtering parameters, such as the applied power and the chamber pressure on the thermal stability has been evaluated for the Ti-Zr-Cu-Pd nanoglass thin-films. The nanoglass thin films revealed higher crystallization temperatures than the chemically comparable melt-spun ribbons. Moreover, it was also observed the enhanced super-cooled liquid region, higher crystallization activation energy, and reduced enthalpy show the higher thermal stability of the nanoglass thin films compared to the melt-spun ribbon. It was concluded that by raising the sputtering pressure, the cluster size reduces and resulted in the increase of the interfacial volume fraction that enhanced the thermal stability of the nanoglass thin films. Here it is revealed that the interfacial regions act as kinetic barriers for the crystal nucleation and growth.

Magnetic

The nanostructured metallic glass shows some interesting magnetic properties compare to the conventional bulk metallic glass of the same chemical composition. The presence of the interfaces and high free volume in the nanoglass affects its magnetic properties due to the different atomic rearrangements and polarization compared to the conventional metallic glass. Ghafari et al., (2012a) have studied the magnetic properties of the FeSc metallic glass and have reported an itinerant constituent in the magnetically ordered state of a FeSc nanoglass at low temperature (≈ 3 K). They present an interesting result of the well-resolved Mössbauer spectroscopy that consists of mainly two sextets, which are ascribed to the core and interfacial regions. One of the resolved sextets has revealed a similar shape and distribution that is perceived in quenched amorphous alloys. Hence, this part is attributed to the core regions of the nanoglass. The additional features in the second resolved part are ascribed to the interfacial regions. It was estimated that the core regions are approx. 58% of the nanoglass and have an internal magnetic field, B_{hf} = 19 T. The interfacial regions are assessed as approx. 34% and the average internal magnetic field is 33 T.

Moreover, evidence of the enhanced magnetic properties of the $Fe_{90}Sc_{10}$ nanoglass compared to the conventional melt-quenched metallic glass of the same chemical composition is exhibited by Witte et al., (2013). The magnetization study was carried out at room temperature that demonstrated the magnetization loops of magnetization *versus* the external magnetic field of a $Fe_{90}Sc_{10}$ nanoglass and conventional ribbon metallic glass of the same composition. It is revealed that the ribbon shows a paramagnetic nature, interestingly, the nanoglass is ferromagnetic with an average magnetization 1.05 μ_B per Fe atom. Authors have analyzed the structures of the nanoglass and melt-spun ribbon using scanning tunneling microscopy and Mössbauer spectroscopy to identify the origin of the ferromagnetism in nanoglass. However, it was not concluded that many other factors influence the ferromagnetism in the nanoglass, but the distinct atomic arrangements at the interface are one reason. Wang et al., (2019) have discussed the inhomogeneous chemical distribution of the nanoglass compared to the conventional melt-quenched metallic glass in the $Sc_{79}Fe_{21}$ system. It was found that the magnetic properties of the nanoglass are different from those of conventional metallic glass. The magnetism was elucidated by the "cluster model," and it was estimated the number of ferromagnetic clusters within the nanoglass is much smaller than that for the conventional metallic glass due to the heterogeneous microstructure of the nanoglass.

Catalytic and Bioactivity

A catalyst is used in different chemical reactions to increase the reaction rate. Most of the catalysts are in the crystalline forms. Nanostructured catalysts are more effective due to their large surface-to-volume ratio resulting in high reactivity (Somwanshi et al., 2020; Mitchell et al., 2021). Metallic nanostructured materials are more popular to act as a catalyst for many chemical reactions (Kwak et al., 2018; Rodrigues et al., 2019; Kodama et al., 2021; Wang et al., 2021). However, there are few studies have been reported on the catalytic activities of nanostructured metallic glass. Bag et al., (2020) have reported the electrochemical detection of glucose using nickel-based nanostructured $Ni_{60}Nb_{40}$ glassy alloys. They have explored the sensing of glucose solutions using different metallic alloys such as nanoglass, nanocompositc, and mclt-spun ribbons of thc samc composition $Ni_{60}Nb_{40}$. It is found that a few nanomolar glucose is detectable using a nanoglass electrode due to high current density. It is assumed that the nanostructure of the nanoglass with glass-glass interfaces and enhanced free volume play an important role in the electrochemical response. The study has shown a detection limit of 100 nM for the nanoglass which is interesting to design a futuristic nonenzymatic glucose sensor as an unconventional, nanostructured, and ligand-free material. Moreover, such remarkable performance result of the nanoglass-based sensor has great potential for the development of diagnostic devices to monitor glucose concentrations with high accuracy.

Chen et al., (2013) have engineered the surface properties of a novel Ti-based nanoglass composite by submicron and nanometer-sized hierarchical glassy structures to tune the cellular behaviors. The experimental results have reported evidence that a new glassy structure is exceptionally advantageous for cell attachments, proliferation, and osteogenic protein or gene expression. The results suggest these hierarchical structures of glassy materials may be extensively applied for surface modifications as a coating on several materials including bulk metallic glass, crystalline metals, oxides, etc. The nanoglass can be considered as a new opportunity for potential applications in a wide variety of technological areas related to biomedical.

Applications

It is implicit from the above discussions that nanostructured metallic glass has exciting properties and needs extensive research to explore further understanding of this type of material. Current activities on the NMG are at the fundamental research level. Hence, any applications at the commercial level are not realized yet. However, there are exciting results of structural, thermal, mechanical, magnetic, electrical, and bioactivity properties that make the material potential for several applications in the area of advanced technologies. The big challenge for a real-time application is the mass production of the NMG, which may take years. It considers that different metallic components of high mechanical properties like improved plasticity and high strength are possible with NMG. These materials will also be very helpful as a catalytic material for many applications such as hydrogen production, glucose detection, fast growth of cells, etc. There are few studies done on the magnetic properties of the NMG and are very interesting. Hence, we may be able to develop new types of magnetic materials that may be useful in the advancement of high-performance magnets, undulators, quantum devices, etc. Similarly, there is huge scope for the discovery of new alloys of immiscible elements in the form of the NMG. It believes that NMG has the potential to bring a next-generation revolution in nanotechnology, and society will benefit from it.

Conclusion

Nanostructured metallic glass is a prospective engineered amorphous material that played a key role in governing and inducing fascinating properties, like ultrahigh stability, superior itinerant magnetism, thermal stability, and enhanced catalytic activity. The interesting properties evolved from the unique atomic packing structures of the core and interface of the NMG. The core atomic structure is similar to the conventional melt-quenched metallic glasses. But, the atomic structure at the interface of NMG is less understood. Hence, it is important to explore the interface atomic structure that could greatly advance our knowledge of the structure-property relationships. This is quite challenging because the glass atomic structure is itself an ambiguous issue for researchers. For a better understanding of the NMG structure, new strategies for synthesis and sophisticated characterization tools need to be adopted to explore how the interfaces modify the properties of NMG. There are several fundamental studies have carried out for understanding of structural-properties relationship of the NMG. However, a limited report is presented on the size effect, thermodynamics, and kinetics of the NMG, hence need more in-depth research. In recent years studies have demonstrated the structural, thermal, mechanical, magnetic, and catalytic properties of the NMG, which are fascinating for potential applications in advanced technologies. No doubt, NMG represents a novel family of amorphous materials that are different from conventional metallic glasses. Hence, NMG will open up new opportunities for their potential use in numerous functional applications.

References

Arnold W, Birringer R, Braun C, Gleiter H, Hahn H, Nandam SH, Singh SP. Elastic moduli of nanoglasses and melt-spun metallic glasses by ultrasonic time-of-flight measurements. *Transactions of the Indian Institute of Metals* (2020) 73:1363-1371.

Bag S, Baksi A, Nandam, S H, Wang D, Ye, X, Ghosh J, Pradeep T, Hahn H. Nonenzymatic glucose sensing using $Ni_{60}Nb_{40}$ nanoglass. *ACS Nano* (2020) 14(5):5543–5552.

Basak AK, Pramanik A, Prakash C, Shankar S, Sehgal SS. Microstructure and micromechanical properties of friction stir processed Al 5086-based surface composite. *Materials Today Communication* (2023) 35:105830.

Bassim N, Scott K, Giannuzzi LA. Recent advances in focused ion beam technology and applications. *MRS Bulletin* (2014) 39:317–325.

Budhani R C, Goel T C, Chopra K L. Melt-spinning technique for preparation of metallic glasses. *Bulletin of Materials Science* (1982) 4:549–561.

Chen N, Shi X, Witte R, Nakayama K S, Ohmura K, Wu H, Takeuchi A, Hahn H, Esashi M, Gleiter H, Inoue A, Louzguinea DV. A novel Ti-based nanoglass composite with submicron–nanometer-sized hierarchical structures to modulate osteoblast behaviors. *Journal of Materials Chemistry B* (2013) 1:2568-2574.

Czichos H, Saito T, Smith L. eds. *Springer Handbook of Materials Measurement Methods*. 2007th ed. Berlin, Heidelberg: Springer; 2006.

Ebrahimi M, Djavanroodi F, Tiji SAN, Gholipour H, Gode C. Experimental investigation of the equal channel forward extrusion process. *Metals (Basel)* (2015) 5:471–483.

Ebrahimi M, Gode C. Severely deformed copper by equal channel angular pressing. *Progress in Natural Science: Materials International* (2017) 27:244–250.

Ghafari M, Hahn H, Gleiter H, Sakurai Y, Itou M, Kamali S. Evidence of itinerant magnetism in a metallic nanoglass. *Appllied Physical Letter* (2012a) 101:243104.

Ghafari M, Kohara S, Hahn H, Gleiter H, Feng T, Witte R, Kamali S. Structural investigations of interfaces in $Fe_{90}Sc_{10}$nanoglasses using high-energy x-ray diffraction. *Appllied Physical Letter* (2012b) 100:133111-1–4.

Giannuzzi LA, Stevie FA eds. *Introduction to Focused Ion Beams*. Boston, MA: Springer; 2005.

Gierak J. Focused ion beam technology and ultimate applications. *Semiconductor Science and Technology* (2009) 24(4):043001.

Gleiter H. Our thoughts are ours, their ends none of our own : Are there ways to synthesize materials beyond the limitations of today? *Acta Mater* (2008) 56:5873–5891.

Gleiter H. Nanoglasses: A New Kind of Noncrystalline Material and the Way to an Age of New Technologies? *Small* (2016) 12:2225–2233.

Guo C, Fang Y, Wu B, Lan S, Peng G, Wang XL, Hahn H, Gleiter H, Feng T. Ni-P nanoglass prepared by multi-phase pulsed electrodeposition. *Materials Research Letters* (2017) 5:293–299.

Halim Q, Mohamed NAN, Rejab MRM, Naim WNWA, Ma Q. Metallic glass properties, processing method and development perspective: a review. *The International Journal of Advanced Manufacturing Technology* (2021)112:1231–1258.

Inoue A, Ohtera K, Kita K, Masumoto T. New amorphous Mg-Ce-Ni alloys with high strength and good ductility. *Japanese Journal of Appllied Physics* (1988) 27:L2248–L2251.

Inoue A, Zhang T, Masumoto T. Al-La-Ni Amorphous alloys with a wide supercooled liquid region. *Materials Transactions JIM* (1989) 30:965–972.

Inoue A, Zhang T, Masumoto T. Zr-Al-Ni amorphous alloys with high glass transition temperature and significant supercooled liquid region. *Materials Transactions JIM* (1990) 31:177–183.

Inoue A, Zhang T, Masumoto T. Glass-forming ability of alloys. *Journal of Non-Crystalline Solids* (1993)156–158:473–480.

Ivanisenko Y, Kübel C, Nandam SH, Wang C, Mu X, Adjaoud O, Albe K, Hahn H. Structure and properties of nanoglasses. *Advanced Engineering Materials* (2018) 20:1–16.

Jiang R, Da Y, Chen Z, Cui X, Han X, Ke H, Liu Y, Chen Y, Deng Y, Hu W. Progress and perspective of metallic glasses for energy conversion and storage. *Advanced Engineering Materials* (2022) 12:1–31.

Katnagallu S, Wu G, Singh SP, Nandam SH, Xia W, Stephenson LT, Gleiter H, Schwaiger R, Hahn H, Herbig M, Raabe D, Gault B, Balachandran S. Nanoglass-nanocrystal composite-a novel material class for enhanced strength-plasticity synergy. *Small* (2020) 16:2004400-1–5.

Klement W, Wiliens RH, Duwez P. Non-crystalline structure in solidified gold-silicon alloys. *Nature* (1960) 187:869–870.

Kodama K, Nagai T, Kuwaki A, Jinnouchi R, Morimoto Y. Challenges in applying highly active Pt-based nanostructured catalysts for oxygen reduction reactions to fuel cell vehicles. *Nature Nanotechnology* (2021) 16:140–147.

Kwak NW, Jeong SJ, Seo HG, Lee S, Kim YJ, Kim JK, Byeon P, Chung SY, Jung WC. In situ synthesis of supported metal nanocatalysts through heterogeneous doping. *Nature Communication* (2018) 9:1–8.

Li J, Doubek G, McMillon-Brown L, Taylor AD. Recent Advances in Metallic Glass Nanostructures: Synthesis Strategies and Electrocatalytic Applications. *Advanced Materials* (2019) 31:1–28.

Lu L, Sui ML, Lu K. Superplastic extensibility of nanocrystalline copper at room temperature. *Science* (2000) 287:1463–1466.

Mao M, Altounian Z, Brüning R. X-ray-diffraction study of structural relaxation in metallic glasses. *Physical Review B* (1995) 51:2798–2804.

Mitchell S, Qin R, Zheng N, Pérez-Ramírez J. Nanoscale engineering of catalytic materials for sustainable technologies. *Nature Nanotechnology* (2021) 16:129–139.

Mohri M, Wang D, Ivanisenko Y, Gleiter H, Hahn H. Thermal stability of the Ti-Zr-Cu-Pd nano-glassy thin films. *Journal of Alloys Compounds* (2018) 735:2197–2204.

Nandam SH, Ivanisenko Y, Schwaiger R, Śniadecki Z, Mu X, Wang D, Chellali R, Boll T, Kilmametov A, Bergfeldt T, Gleiter H, Hahn H. Cu-Zr nanoglasses: atomic structure, thermal stability and indentation properties. *Acta Materials* (2017) 136:181–189.

Natter H, Schmelzer M, Hempelmann R. Nanocrystalline nickel and nickel-copper alloys: Synthesis, characterization, and thermal stability. *Journal of Materials Research* (1998) 13:1186–1197.

Oliver WC, Pharr GM. Measurement of hardness and elastic modulus by instrumented indentation: Advances in understanding and refinements to methodology. *Journal of Materials Research* (2004) 19:3–20.

Phaneuf MW. Applications of focused ion beam microscopy to materials science specimens. *Micron* (1999) 30:277–288.

Rodrigues TS, Da Silva AGM, Camargo PHC. Nanocatalysis by noble metal nanoparticles: controlled synthesis for the optimization and understanding of activities. *Journal of Materials Chemistry A* (2019) 7:5857–5874.

Schroers J. Processing of bulk metallic glass. *Advanced Materials* (2010) 22:1566–1597.

Sharma A, Nandam SH, Hahn H, Prasad KE. Effect of structural relaxation on the indentation size effect and deformation behavior of Cu-Zr-based nanoglasses. *Frontiers in Materials* (2021) 8:676764.

Singh SP, Chellali MR, Velasco L, Ivanisenko Y, Boltynjuk E, Gleiter H, Hahn H. Deformation-induced atomic rearrangements and crystallization in the shear bands of a $Tb_{75}Fe_{25}$ nanoglass. *Journal of Alloys Compounds* (2020a) 821:153486-1–8.

Singh SP, Witte R, Clemens O, Sarkar A, Velasco L, Kruk R, Hahn H. Magnetic $Tb_{75}Fe_{25}$ nanoglass particles for cryogenic permanent magnet undulator. *ACS Appllied Nano Materials* (2020b) 3:7281–7290.

Somwanshi SB, Somvanshi SB, Kharat PB. Nanocatalyst: A brief review on synthesis to applications. *Journal of Physics: Conference Series* (2020) 1644:012046.

Tabbakh T, Kurdi A, Basak AK. Effect of strain rate and extrinsic size effect on micro-mechanical properties of Zr-based bulk metallic glass. *Metals (Basel)* (2021) 11(10):1611.

Tong Y, Dmowski W, Witczak Z, Chuang CP, Egami T. Residual elastic strain induced by equal channel angular pressing on bulk metallic glasses. *Acta Materials* (2013) 61:1204–1209.

Valiev RZ, Langdon TG. Principles of equal-channel angular pressing as a processing tool for grain refinement. *Progress in Materials Science* (2006) 51:881–981.

Volkert CA, Minor AM. Focused ion beam micromachining. *MRS Bulletin* (2007) 32:389–399.

Wang C, Palit M, Yin N, Shi Q, Ivanisenko Y, Gleiter H, Hahn H. Magnetic contributions to the low-temperature specific heat of $Sc_{79}Fe_{21}$ nanoglass. *Journal of Appllied Physics* (2019) 125:045111.

Wang JQ, Chen N, Liu P, Wang Z, Louzguine-Luzgin DV, Chen MW, Perepezko JH. The ultrastable kinetic behavior of an Au-based nanoglass. *Acta Materials* (2014) 79:30–36.

Wang N, Sun Q, Zhang T, Mayoral A, Li L, Zhou X, Xu J, Zhang P, Yu J. Impregnating subnanometer metallic nanocatalysts into self-pillared zeolite nanosheets. *Journal of the American Chemical Society* (2021) 143:6905–6914.

Wang XD, Cao QP, Jiang JZ, Franz H, Schroers J, Valiev RZ, Ivanisenko Y, Gleiter H, Fecht HJ. Atomic-level structural modifications induced by severe plastic shear deformation in bulk metallic glasses. *Scripta Materials* (2011) 64:81–84.

Wang YB, Qu DD, Wang XH, Cao Y, Liao XZ, Kawasaki M, Ringer SP, Shan ZW, Langdon TG, Shen J. Introducing a strain-hardening capability to improve the ductility of bulk metallic glasses via severe plastic deformation. *Acta Materials* (2012) 60:253–260.

Witte R, Feng T, Fang JX, Fischer A, Ghafari M, Kruk R, Brand RA, Wang D, Hahn H, Gleiter H. Evidence for enhanced ferromagnetism in an iron-based nanoglass. *Applied Physics Letter* (2013) 103:073106.

Yavari AR, Moulec A, Inoue A., Nishiyama N, Lupu N, Matsubara E, Botta WJ, Vaughan G, Michiel MD, Kvick A. Excess free volume in metallic glasses measured by X-ray diffraction. *Acta Materials* (2005) 53:1611–1619.

Zhilyaev AP, Langdon TG. Using high-pressure torsion for metal processing: Fundamentals and applications. *Progress in Materials Science* (2008) 53:893–979.

Chapter 5

Exploring Bulk Metallic Glasses Using Computational Methods

Modalavalasa Kishor[1]
Kunal Chopra[2]
Shrey Dixit[3]
G. Jaideep Reddy[1]
and A. K. Prasada Rao[1,*]
[1]BML Munjal University, India
[2]Faculty of Computer Science, TU Dresden, Germany
[3]The University of Hamburg, Germany

Abstract

Over the past several decades bulk metallic glasses have gained special recognition owing to their exceptional hardness, wear resistance, high Young's modulus, and corrosion resistance. There are several theories which explain their formation on an atomic scale. The ease of formation of metallic glassy structure has been measured in terms of glass forming ability. Further, the glass forming ability depends on various thermodynamic and topological parameters. However, it is always a difficult challenge to arrive at a bulk metallic glassy alloy with optimized chemical composition through experimental methods. Therefore, data driven computational approaches have become more powerful tools in designing optimum alloys. These approaches include artificial intelligence and machine learning methods. This chapter critically reviews the historical evolution of such methods used in development of

---

* Corresponding Author's Email: akprasada@yahoo.com.

In: Properties and Uses of Metallic Glass
Editor: Shiv Prakash Singh
ISBN: 979-8-89113-569-7

bulk metallic glasses and their modal architecture. Finally best AI/ML models have been recommended herein.

Keywords: glass forming ability, critical thickness prediction, machine learning models, neural networks

Introduction

Bulk metallic glasses and traditional amorphous alloys differ in the processes of their formations, with the former being produced at very low critical cooling rates and the latter requiring high cooling rates to suppress the nucleation of crystalline phases. The limited range of possible structures that can be prepared using traditional glassy alloys, such as powders, films, and ribbons, contrasts with the unique properties of BMGs, which exhibit amorphicity, high strength, and a glass transition. This transition transforms supercooled liquids into a glassy state when cooled from high to low temperatures, and vice-versa, causing abrupt changes in thermodynamic and physical properties at the glass transition temperature (T_g).

Research on metallic glasses (MG) is closely linked to that on metallic liquids, as both have disordered atomic configurations. To retain the non-crystalline structure, liquids can be quenched at high cooling rates, estimated to be on the order of 10^5~10^6 K s^{-1} for the alloys. The first glassy alloy was produced in 1959 by a research group at Caltech using a splat quenching technique on a composition of $Au_{75}Si_{25}$ (Klement et al. 1960). This pioneering work was followed by the discovery of many metallic glasses in various alloy systems, thanks to the invention of melt spinning. These new alloys offer the potential for a wide range of applications, and ongoing research aims to expand the range of bulk forms that can be produced using bulk metallic glasses (BMGs).

For around two decades after the first manufactured metallic glass (Klement et al. 1960), only thin foils of metallic ribbons were formed as rapid quenching was required with a cooling rate greater than 10^6 K/s to avoid crystallization and freeze atoms in a random, non-periodic fashion which gave metals low heat conductivity. However, we will be discussing glasses with diameters at least 1 mm consisting of three or more alloying elements manufactured at a relatively low solidification rate of less than 10^3 Ks^{-1}. To achieve the thickness of 1 mm, various methods were been deployed starting from drop-tube processing of Pd-Si-Cu by solidifying the melt in a containerless low-gravity environment providing low surface access to the alloy for heterogeneous nucleation (Steinberg et al. 1981). Drehman et al. (1982) tried thermal cycling (successive heating and cooling) to avoid heterogeneous nucleation in the Pd-Ni-P alloy system. Then, Xu et al. (1990) cooled the melt under high pressure in a belt apparatus which suppressed the long-range atomic motion, reducing the crystallization despite the presence of nucleation centers. No doubt, 80 cm size has been achieved in Palladium (Nishiyama et al. 2012) based amorphous alloy, but it is limited to small volumes due to the use of expensive Palladium which limited the entry of industries into BMG research. Another important property to be considered here about BMGs is the requirement of a large, supercooled region i.e., the difference between its glass transition and crystallization temperatures which Inoue et al.(1989) proved to be 70 K ΔT_x in the Al-La-Ni alloy system. It was followed by Inoue et al. (1990) in a ternary alloy system (Zr-Al-Ni) through a melt-spinning technique which went up to 77 K for Zr rich alloys. It was until then limited to highly expensive alloys such as 65 K for Pd-Ni-P and 70 K for Pt-Ni-P (Chen and Jackson, 1978). The wide supercooled region along with high T_g/T_m ratio provided a greater stability region for resistance to nucleation leading to a higher glass forming ability (GFA) and a slow cooling rate up to 100 K/s which eased the production of BMGs. Inoue et al. (1989) used water quenched to form an $Al_{25}La_{50}Ni_{25}$ alloy cylinder of diameter 1.2 mm. So, determining the size of the glass formed depends mostly on the cooling rate of the process.

Although the formation of BMGs is highly influenced by its GFA, the effect of casting parameters can't be ignored. Laws et al. 2009 demonstrated that the heat transfer coefficient influences the thickness of the glassy zone. Casting parameters include casting velocity, melt superheat (casting temperature), mold thickness and roughness. Recently, non-conventional methods were used to develop BMGs with more intricate geometries, which could not be possible through traditional deformation/ machining methods due to the high hardness and brittle nature of BMGs, Non-conventional methods include ultrasonic layer-by-layer additive manufacturing process (Wu et al. 2019), welding through a pulsed laser beam (Shao L et al. 2017), spark plasma sintering technology (Wang et al. 2013), pneumatic injection additive manufacture (Wu W et al. 2018) and laser additive manufacturing (Williams E et al. 2017). Within laser additive manufacturing, Selective Laser Melting (SLM) is a much-preferred technique due to the rapid cooling involved and providing more flexibility with the geometries.

GFA of an alloy liquid is its ability to form a glassy phase during solidification. Achieving a high GFA has traditionally been the basis for designing new BMGs. Since the discovery of BMGs, multiple criteria for determining GFA have been employed, such as finding the maximum thickness, R_c was difficult to obtain experimentally. However, the majority of those criteria were based on the transformation temperatures of alloys majorly, T_g (Glass transition temperature), T_x (Crystallization temperature) and T_l (Liquidus temperature). BMG formation requires a wide distribution of atomic sizes which can be achieved by undercooling the melt providing resistance to the kinetic energy during glass formation. Based on this phenomenon, some of the criteria developed for predicting GFA included T_{rg} (Reduced glass transition temperature) which is the ratio of T_g to T_l (Turnbull, 1969). While $T_{rg} > 2/3$ provided a good starting point for maintaining ease in manufacturing BMG but it said nothing about the stability after its formation. The stability condition was provided by (Inoue, 1993) through the difference of T_x and T_g as ΔT_x. ΔT_x is independent of the technique used for glass forming and is based on the thermal stability of the glass after it has been formed. To combine the features of both the above parameters, Lu and Liu (2002) proposed the gamma parameter which is $T_x/(T_g+T_l)$. (Suryanarayana et al. 2009) concluded that not only the stability of glass (ΔT_x) but measurement of ease of glass formation through T_{rg} is equally important in determining the GFA. Also, Kim et al. (2005) provided a strong correlation between T_{rg} and GFA based on experimental results of BMG obtained from different compositions and they also validated T_{rx} (reduced crystallization temperature) which has been used instead of T_g due to their similarity for metallic glasses and considered both the resisting and stability criteria of the alloys but showed a better correlation with R_c when mapped linearly to the experimental data.

The glass-forming ability (GFA) of BMG alloys was found to be strongly dependent on their composition, good glass formers usually have compositions close to the eutectic points. In recent years, multicomponent glass formers have been developed that exhibit excellent glass-forming ability and low critical cooling rates, without requiring noble metals as essential constituents. Several research works reported that achieving high GFA through the addition of alloying elements like Fe (Makino et al. 2007), Yttrium (Malekan et al. 2023), Cu (Kim et al. 2003), Zr (Kuhn et al. 2002), Ti (Hofmann et al. 2008), Mg (H. Ma et al. 2003) and La (H. Tan et al. 2002). (Lee et al. 2011) were able to improve both GFA and plasticity by improving the atomic packing state and mixing enthalpy in a Cu-Zr based BMG. So, selecting the alloy is an important parameter in improving GFA.

Data Collection and Augmentation

Data is a crucial prerequisite for computational approaches and serves as the sole source for enhancing the efficiency of the models. This data can be obtained through simulations, specifically Density Functional Theory (DFT), or from experimental studies. Beginning in 2004, computational model advancements commenced with a limited number of data points, primarily focusing on concise data related to binary and ternary alloys. However, models like artificial neural networks (ANN) and deep learning necessitate larger amounts of data for effective operation. The Landolt-Bornstein handbook was developed with a huge source of data points on non-crystalline materials and their glass forming ability. This handbook enabled the list of machine-learning models to predict the glass forming ability. Later, Long et al. (2009) developed a vast dataset comprising temperature descriptors such as T_g (glass transition temperature), T_x (onset crystallization temperature), and T_l (liquid temperature). This dataset encompasses a wide range of elements, including Cu, Ca, Mg, La, Pd, Ti, Pr, Y, Co, Au, Hf, Gd, Zr, Fe, and Ni. Over the last decade, this data has paved the way for the development of thermodynamic descriptors by establishing interrelationships among T_g, T_l, and T_x. Xiaodi et al. (2020) developed a back propagation neural network model with 3277 ternary alloys for the prediction of MG's and non-MG's. Similarly, Logan Ward et al. (2016) compiled a dataset comprising the range of supercooled liquids (621), critical casting thickness (5916), and glass-forming ability (6315). This data was collected from 41 independent research works as well as a databook. Over the past few years, researchers have utilized this freely available data in various formats, employing clustering techniques in computational studies. In the past ten years, there has been significant progress in the development of computational analytics to study the dynamics of virtual BMG. While CALPAD analysis was originally introduced in 2004, its application was initially limited to ternary and quaternary alloys. This restriction was due to the challenges posed by solving higher numbers of alloying additions and the vast compositional space. Figure 1 illustrates the data on publications during the last decade and shows an enhanced trend in computational research from 2013 to 2023.

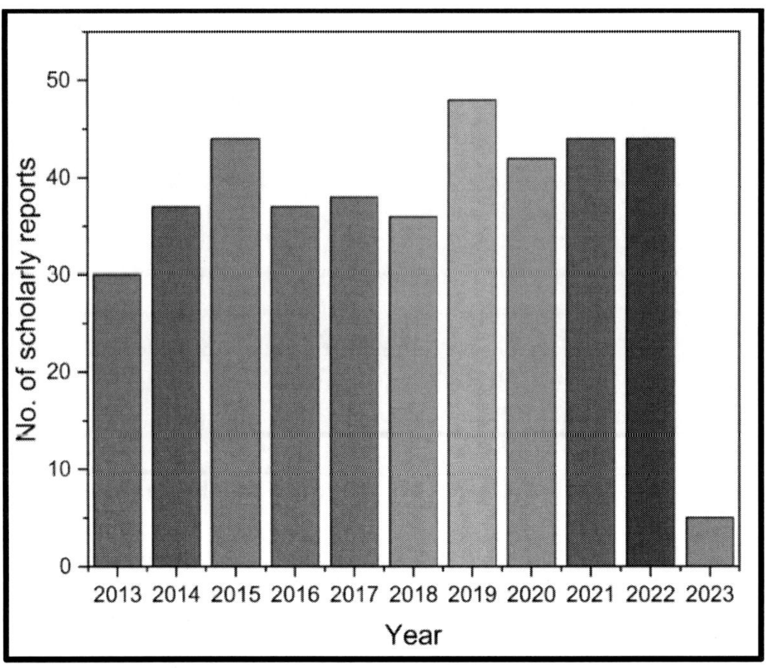

Figure 1. Articles published on the computational approaches of BMGs.

Evolution of Thermodynamic Descriptors

Since the thermodynamics properties belong to states above or below the T_g, fragility was introduced to indicate the degree of viscosity deviation. We know that undercooling of a liquid can be accelerated by increasing the viscosity but only after constraining it kinetically (Angell, 1995). building on this knowledge, G. J. Fan et al. (2005) found out that M, the kinetic fragility of a glass-forming liquid is positively co-related to $\Delta S_f/(T_m-T_k)$, where ΔS_f (entropy of fusion), T_k - (Kauzmann temperature) and T_m (melting point) giving thermodynamic fragility as an indicator for predicting the glass forming ability as a higher m value corresponds to a more fragile liquid. Also, Xuelian Li et al. (2007) expanding on the strong and fragile liquids, through the results of 4 different glass forming systems, found a linear relationship between $|\Delta H^{chem}|$(Mixing Enthalpy) and T_g. It was attributed to the fact that an increase in both $|\Delta H^{chem}|$ and T_g is possible due to the change in interaction intensity between atoms of different compositions. Xuelian Li et al. (2016) proposed the parameter of $(S_\sigma/k_B)/M$ as a predictor of GFA wherein a lower value of M gives lower energy and a stable liquid while increasing the chemical entropy which is the mismatch entropy (S_σ/k_B) will increase the atomic diffusivity giving a better GFA. The authors found its value to be greater than 0.17 to form BMGs and a good differentiator between BMGs and Marginal Metallic Glasses (MMGs) when M =1 which is usually the case for Metallic Glasses (Yogesh Prabhu, 2020). Many authors then tested the GFA parameter P_{HS}, product of $|\Delta H^{chem}|$ and (S_σ/k_B), combining both thermodynamic and topological parameters on different compositions, Zr-Cu-Al-Ni (27), Cu-Zr-Ag-Hf (28), Cu-Zr-Ti-Ni (29), Zr-Ti-Cu-Ni-Be (30), and Zr-Ti-Cu-Ni-Al (31) and found a strong negative correlation with GFA. While thermal parameters can predict a good value of GFA, we need to consider the structural and topological features of the alloy as well. Egami (T. Egami, 2002) through the theory of increasing glass transition temperature by a repulsion between small metallic atoms had given rules as a) high atomic size ratio, b) higher number of elements for a single BMG, c) high interaction between smaller and larger atoms and d) artificially induced repulsive interaction among smaller atoms. Inoue (Takeuchi et al. 2005) Classification of bulk metallic glasses by atomic size difference, heat of mixing and period of constituent elements and its application to characterization of the main alloying element. Miracle and Senkov (Miracle and Senkov, 2003) provided a topological model based on the ratio of the radius of solute vs solvent elements which if has a value between 0.5 and 1 decreases the critical concentration destabilizing the host lattice forming metallic glasses. (Mukherjee et al. 2004) described a high GFA for alloys that have high viscosity at the liquidus temperature as it reduces the volume change hence resulting in reduced crystallization of the alloys.

Computational Analysis of BMGs

Developing high-strength, high-hardness BMGs for contemporary applications is a challenging task, primarily due to the vast number of potential compositions for each alloy. Given the significant time and cost involved in the experimental process of discovering an optimized property-oriented BMG, computational approaches have been developed to address this issue. These computational approaches aim to reduce costs and save time by utilizing thermal parameters and experimental data to search for optimized BMGs. Initial computational approaches were developed using CALPAD; later machine learning techniques enhanced the prediction scope. CALPAD calculations promoted the prediction possibilities of GFA and developed new thermodynamic descriptors for optimizing D_{max} (critical casting diameter). Deok et al. (2004) introduced a new empirical rule known as the driving force criterion, which helps in predicting the optimal composition for maximizing GFA using CALPAD calculations. The researchers calculated the driving forces for all the crystalline phases in the Cu-Ti-Zr test alloy and observed the presence of two local minima. Interestingly, the obtained local minimum values maintained a ratio of 3:1 and 1:3 for Zr:Ti. Notably, these ratios closely matched the compositions proposed by Inoue and Johnson. This finding suggests a strong agreement between the driving force criterion and the previously identified compositions that exhibit high GFA. Correspondingly, the minimum driving force criterion was applied to Al-Cu-Zr and Mg-Cu-Y alloys, yielding results that demonstrated strong agreement with experimental validations (Bo et al. 2010; Kim et al. 2005). Jitendra Gaur et al. (2016) analyzed three different Pd-based bulk metallic glasses (BMGs) using Taylor series expansion. The focus was on investigating thermodynamic properties such as Gibbs phase energy, entropy difference, and enthalpy difference. The theoretically calculated thermodynamic descriptors were found to be in accordance with experimental results. Binary plots were generated to depict the relationships between these descriptors and the temperature difference (ΔT) within the temperature range spanning from T_m to T_g. Han et al. (2017) developed a CALPAD approach for the quantification of compositional design protocol (CDP), aimed at achieving high GFA using thermodynamic descriptors. In this study, the researchers took into account the atomic structure and local composition to determine the positive or negative impact of different phases on the GFA of Fe-based alloys. Consequently, the study also provided a set of guidelines for element selection based on thermodynamic descriptors for effective phase selection. Machine learning predictions of GFA were implemented to improve the accuracy of predictions. In comparison, the CALPHAD approach has few limitations in terms of data and model complexities. Consequently, this has facilitated the development of machine learning models to address complex studies with a reduced number of data points. K. G. Keong et al. (2004) developed artificial neural network models for predicting the crystallization temperatures of Ni-P amorphous alloys. The study described how the heating rate and phosphorus content affect the crystallization peak temperatures of the Ni-P alloys.

Cai et al. (2010) designed a radial basis function artificial neural network (RBFANN) model to predict the undercooling degree (ΔT_x). The study considered the undercooling degree data of La-Al-Ni ternary alloys, including Cu, La, Fe, Zr, and Mg. Eleven relevant thermodynamic descriptors were employed to predict the ΔT_x, and the model achieved an R value of 0.9862. Using the RBF ANN model, a new set of Zr-Al-Ni-Cu bulk metallic glasses (BMGs) was developed, and there ΔT_x was predicted. In addition to predicting the undercooling degree (ΔT_x), the same model was also utilized to predict the glass transition temperature (T_{rg}), achieving an R value of 0.9672. This indicates a strong correlation between the predicted Trg values and the actual values (A. H. Cai et al. 2008).

Majid et al. (2015) conducted a subsequent study to address the complex task of predicting the critical casting thickness (D_{max}) for glass-forming alloys (GFAs). Due to the non-linear behavior of D_{max} with respect to thermal properties, the researchers explored multiple algorithms including Support Vector Machine (SVM), Artificial Neural Networks (ANN), Multiple Linear Regression (MLR), General Regression Neural Network (GRNN), and computational techniques. The study utilized thermal descriptors such as glass transition temperature, crystallization temperature, and liquidus temperature (T_l), which were found to be strongly correlated with the GFA criteria. However, despite their efforts, the study did not achieve a higher level of prediction accuracy. Interestingly, the Computational Intelligence (CI) models outperformed the statistical and physical models in terms of performance.

Manwendra et al. (2016) predicted the glass-forming ability using genetic programming and developed new parameters, G_p, to estimate the GFA. Y. T. Sun et al. (2017) developed machine learning algorithms for better accuracy in predicting GFA. This study developed an SVM model with binary test data and was able to predict new MGs (metallic glasses). It also demonstrated the relationship and relevance of ΔT_l descriptors in GFA prediction. Logan Ward et al. (2018) designed ML models to predict three different properties: GFA, D_{max}, and supercooled liquid range (ΔT_x). For predicting the D_{max} values of the metallic glasses (MGs), a dataset of 5916 data points was collected. Ten ML algorithms, including RF, REPTrees, and several ensemble techniques, were developed and evaluated for performance using 10-fold cross-validation and leave-binary-out tests. Both methods performed well and were able to increase the overall model efficiency. A correlational study was conducted by Jianqing et al. (2021) to examine the relationship between elastic modulus and GFA in the ZrCuAl(X) amorphous system, employing a machine learning approach. The study utilized a support vector machine (SVM) model to investigate the impact of Ni, Nb, and Nb additions (ranging from 1% to 6%) on the base alloy. As for the network architecture, the authors employed Support Vector Regression with the Pearson universal kernel (PUK) and radial basis function (RBF) kernel, although the details provided in the text are somewhat contradictory to the accompanying figure. The figure depicts what appears to be several decision trees (random forests) whose outputs are subsequently fed into an SVR, whereas the textual explanation lacks any mention of this. Curiously, the authors opted for an unconventional evaluation metric. The reason behind their choice to deviate from more common metrics like RMSE or R2 remains unexplained. Overall, the lack of coherence between the textual description and the figure renders the above work less trustworthy.

Daniel et al. (2021) utilized a dataset comprising 480 Fe-based BMG alloys collected from literature, with varying thicknesses ranging from 0.055 to 18 mm to develop machine learning models for specific Fe-based BMGs. The performance evaluation on a validation dataset resulted in an R^2 value of 0.71. The authors employed two models: multiple linear regression and tree-boosting and justified the use of both models by stating that multiple linear regression offers interpretability, while tree-boosting can capture complex relationships and exhibit higher predictive power. However, combining the predictions of both models by taking their mean introduces a compromise, as it sacrifices interpretability and may lead to suboptimal results by including the linear regression component. Instead, the employment of techniques in machine learning, such as Shapley Analysis (Lundberg et al. 2017) or LIME (Ribeiro et al. 2016) on tree-boosting results to gain insights into the model's decision-making process provides better interpretability. Also, regularization techniques could have been employed to avoid overfitting in the XGBoost model and potentially improved the R^2 value of their model as it contributed to a lower performance on the validation data for the tree model.

Samavatian et al. (2021) used a dataset with 7950 alloy compositions to employ a correlative-based neural network to design novel quaternary BMGs. The study demonstrated that quaternary ZrCoAlNi alloys exhibited high glass-forming ability within a wider composition range. The network architecture employed in this study was a novel correlation-based neural network design. The data underwent a processing stage involving the multiplication of pairs of variables by pretrained correlation matrices, followed by a ReLU activation layer, and average pooling. Lastly, the processed data was flattened and concatenated with the unprocessed data. This concatenated input was then fed into a fully connected neural network with two hidden layers. The performance evaluation of the model resulted in an RMSE of 0.6159 for D_{max} and 0.00012645 for T_{rg}. The authors mentioned the classification of the dataset to indicate the alloys' ability to form BMG, MG ribbon, or crystalline type, but the specific methodology for this classification was not provided, posing challenges to the result reproducibility. Furthermore, they acknowledged using average values for unique alloy compositions with conflicting reports in the literature, which raises concerns regarding the validity of including such alloys in the analysis. A noteworthy aspect of their work is the unique validation procedure, where actual BMGs were produced and used to assess the model's performance.

Xiong et al. (2021) did a classification study of 695 alloys to differentiate between L-GFA (limited glass forming ability having 1mm < D_{max} < 5mm and G-GFA (good glass forming ability) having D_{max} > 5mm. Then, for G-GFA alloys, the specific D_{max} value was predicted. The authors experimented with various models but ultimately selected XGBoost for both classification and regression tasks based on cross-validation accuracy. The performance evaluation on the validation dataset resulted in an R^2 value of 0.6419. Notably, the authors conducted feature importance calculations using XGBoost models, yielding interesting findings. The analysis revealed that R_m, ΔH_f, and S_{mix} were the most influential factors in determining the GFA category (higher or lower), whereas ΔH_f emerged as the most important predictor for determining the precise D_{max} value. These insights highlight the significance of specific alloy characteristics in differentiating GFA levels and accurately estimating D_{max} in G-GFA alloys.

Reddy, G et al. (2021) collected data on 400 BMGs with Cu, La, Ca, Mg, Fe, Zr, Ti, Pd, Co, Pr, Hf, Gd, Au, Y, and M as major constituent elements. The primary prediction task of the study was to estimate the D_{max} of these BMGs. The network architecture employed was a feedforward neural network. The performance evaluation on the test dataset resulted in an R-squared value of 0.71. A noteworthy aspect of this work is the consideration of the proportion of small, medium, and large elements in the alloy based on the work of Inoue et al. (2021) The authors were motivated by the *radial range theory of classification* (Reddy et al. 2021), which posits that in a multi-component BMG system, the atomic radii difference should exceed 12%. By incorporating this information into their input data, the authors observed a significant contribution to their prediction performance. This inclusion of the relative proportions of different-sized elements adds an interesting dimension to the study, aligning with the underlying principles of atomic structure and its impact on BMG properties.

Ziqing et al. (2023) developed an unsupervised generative adversarial network algorithm to generate diverse compositions of Bulk Metallic Glasses (BMGs). The algorithm demonstrated excellent performance, particularly with small datasets, and showcased its capability for inverse design of BMGs with specific compositions. Long et al. (2023) designed the Deep Forest algorithm based on Random Forest for predicting D_{max}. The developed model provides prediction errors of less than +5 when using optimal Thermodynamic descriptors. It demonstrates higher accuracy (R2 = 0.763) compared to other mainstream models, such as SVR, RF, GBDT, KNN, and XG-Boost. Shapley analysis was utilized to understand feature importance, and the results indicated that the variable φ ($\varphi = \alpha + S_{mix}/2$) holds greater significance in predicting Bulk Metallic Glasses (BMGs). Similarly, Long et al. (2023) developed two different models known as Improved Deep Neural, Network (IDNN) and MSEDNN, both utilizing the dense Loss function. A dataset of 700 Bulk Metallic Glasses (BMGs) was collected for the prediction, and a generalization study was conducted. The MSEDNN model outperformed the IDNN model, achieving higher R2 values of 0.841 and 0.781, respectively. Additionally, Shapley points were generated for various thermodynamic descriptors to understand their importance in the models. Ravindranath et al. (2023) conducted a study that investigated various algorithms, including Random Forest (RF), XG-Boost (XGB), K-Nearest Neighbours (KNN), Support Vector Machine (SVM), Gradient Boosting Decision Tree (GBDT), Decision Tree, and Artificial Neural Networks. Among these algorithms, RF and XGB outperformed all the others in predicting Glass-Forming Ability (GFA). Recently, Liu, G et al. (2023) demonstrated and emphasized the significance of incorporating physical insights into predicting the Glass Forming Ability (GFA) using Machine Learning techniques. The authors employed the same machine learning architecture to predict GFA, but they trained it on four distinct types of data and subsequently compared the performance of these four models. The first model utilized 201 general-material features derived from 31 elemental features. The second model employed random unphysical features. The third model, in contrast, solely utilized the alloy composition and excluded all other features. Finally, the fourth model exclusively relied on three features termed "Human Learning Features." These features, namely Liquidus temperature reduction ΔT, Atomic size difference δ, and Maximum heat of mixing ΔH_{mix}, were developed through years of research in the BMG literature. Even before the introduction of Machine Learning techniques to this field, these features were successfully utilized for GFA prediction in BMGs, representing state-of-the-art knowledge in the domain. Following the training process, the authors made an intriguing discovery: the fourth model trained on Human Learning Features significantly outperformed the other three models when tested on new, previously unseen data. This finding demonstrates that machine learning models trained on Human Learning Features not only excel in generalizing well, on unseen data but also have broader applications in identifying potential BMG systems. This highlights the crucial role of domain knowledge as an essential and non-negotiable component of data science. By integrating physical insights and domain expertise into the Machine Learning process, researchers can achieve more accurate predictions and enhance their understanding of complex systems like BMGs.

Conclusion

The literature confirms the existence of extensive research studies on bulk metallic glasses (BMGs), which have shown an increased probability of obtaining superior materials for various applications. The availability of an infinite compositional space has opened numerous new possibilities for designing and developing extraordinary materials. Computational approaches have facilitated this process by reducing both time and cost requirements. Over the past two decades, computational models such as machine learning and CALPAD calculations have significantly improved in terms of efficiency and accuracy, simplifying the work of experimentalists in designing new materials. Machine learning models have played a crucial role in predicting the critical casting diameter (D_{max}) and properties of virtual BMGs. This study provides a comprehensive discussion of the computational approaches developed for predicting the critical casting diameter and properties. Detailed discussions focus on machine learning prediction models as well as the challenges and opportunities for future development.

Recently, computational approaches, such as the Radius Range theory demonstrated in the authors' work, have helped clarify the ambiguity in atomic size classification that persisted until recent times. These computational methods have gained importance as they accelerate research in exploring the glass forming ability, mechanical properties, and optimal chemical composition of unknown BMGs without the need for extensive experiments. Computational approaches such as CALPHAD also have limitations as they are based on certain assumptions, and hence deviate from experimental results.

Similar to other machine learning applications, predicting the GFA of BMGs is also highly reliant on data. The quantity and quality of available data play a crucial role in determining the accuracy of predictions, regardless of the chosen predictive model. In essence, Liu, G et al. (2023) emphasizes the crucial relationship between physical insights and Machine Learning for predicting GFA in BMGs. The successful implementation of domain knowledge through the Human Learning Features opens up exciting possibilities for uncovering and discovering new BMG systems with profound implications for various applications. A further challenge arises from the fact that various research papers employ different evaluation metrics, making it arduous to compare and contrast different approaches effectively. Additionally, when the code for a particular approach is not open source, it becomes exceedingly difficult to replicate results and conduct meaningful comparisons. Among various machine learning techniques, Decision Tree-based models, such as Random Forests and XG_{Boost}, have shown commendable performance in scenarios with limited or abundant data. However, in cases where ample data is available, neural networks might outperform other models. Therefore, it is crucial to experiment with several techniques before settling on one for a specific application. Moreover, interpreting the model outputs is essential to understanding why certain features influence the GFA of BMGs. Techniques like Shapley Analysis and LIME enable us to interpret these black-box models, potentially leading to improved prediction performance. Recent advances in generative artificial intelligence show promise with improved accuracy compared to conventional methods, offering potential future improvements in exploring optimum alloys for challenging applications.

References

Angell CA. Formation of glasses from liquids and biopolymers. *Science* (1995) 267(5206):1924-1935.

Bhatt J, Dey GK, Murty BS. Thermodynamic and topological modeling and synthesis of Cu-Zr-Ti-Ni–based bulk metallic glasses by mechanical alloying. *Metallurgical and Materials Transactions A* (2007) 39(7):1543-1551.

Bhatt J, Murty BS. Thermodynamic modeling of Zr-Ti-Cu-Ni-Be Bulk metallic glass. *Transactions of the Indian Institute of Metals* (2009) 62(4-5):413-416.

Bo H, Wang J, Jin S, Qi HY, Yuan XL, Liu LB and Jin ZP. Thermodynamic analysis of the Al-Cu-Zr bulk metallic glass system. *Intermetallics* (2010) 18(12):2322–2327.

Bobbili, R. Interpretable glass forming ability prediction of amorphous alloys through tree based algorithms. *Materials Letters* (2023) 349: 134774.

Cai AH, Xiong X, Liu Y, An WK, Tan JY. Artificial neural network modeling of reduced glass transition temperature of glass forming alloys. *Applied Physics Letters* (2008) 92:111909.

Cai AH, Xiong X, Liu Y, An WK, Tan, JY, Luo Y. Artificial neural network modeling for undercooled liquid region of glass forming alloys. *Computational Materials Science* (2010) 48(1): 109–114.

Chen HS. Metallic glasses. *Materials Science and Engineering* (1976) 25:59-69.

Drehman AJ, Greer AL, Turnbull D. Bulk Formation of a metallic glass: $Pd_{40}Ni_{40}P_{20}$. *Applied Physics Letters* (1982) 41(8):716-717.

Fan GJ, Choo H, Liaw PK. Fragility of metallic glass-forming liquids: A simple thermodynamic connection. *Journal of Non-Crystalline Solids* (2005) 351(52-54):3879-3883.

Gaur J, Mishra RK. Analysis of thermodynamic properties for Pd-based bulk metallic glasses. *Journal of Alloys and Compounds* (2016) 658:465–469.

Han JJ, Wang CP, Wang J, Liu XJ, Wang Y, Liu ZK. Compositional design of Fe-based multi-component bulk metallic glass based on CALPHAD method. *Materials and Design* (2017) 126:47–56.

Hofmann DC, Suh JY, Wiest A, Duan G, Lind ML, Demetriou MD and Johnson WL. Designing metallic glass matrix composites with high toughness and tensile ductility. *Nature* (2008) 451(7182):1085-1089.

Inoue A, Zhang T, Masumoto T. Zr-Al-Ni amorphous alloys with high glass transition temperature and significant supercooled liquid region. *Materials Transactions, JIM* (1990) 31(3):177-183.

Inoue A, Zhang T, Masumoto T. Al-LA-Ni amorphous alloys with a wide supercooled liquid region. *Materials Transactions, JIM* (1989) 30(12):965-972.

Inoue A, Kita K, Zhang T, Masumoto T. An amorphous $LA_{55}Al_{25}Ni_{20}$ alloy prepared by water quenching. *Materials Transactions, JIM* (1989) 30(9):722-725.

Inoue A, Zhang T, Masumoto T. Glass-forming ability of alloys. *Journal of Non-Crystalline Solids* (1993) 156-158:473-480.

Klement W, Willens RH, Duwez P. Non-crystalline structure in solidified Gold–Silicon Alloys. *Nature* (1960) 187(4740):869-870.

Kim YC, Kim DH, Lee JC. Formation of ductile Cu-based bulk metallic glass matrix composite by Ta addition. *Materials Transactions* (2003) 44(10):2224-2227.

Keong KG, Sha W, Malinov S. Artificial neural network modelling of crystallization temperatures of the Ni-P based amorphous alloys. *Materials Science and Engineering A* (2004) 365(1–2):212–218.

Kim D, Lee BJ, Kim NJ. Thermodynamic approach for predicting the glass forming ability of amorphous alloys. *Intermetallics* (2004) 12(10-11):1103–1107.

Kim D, Lee BJ, Kim NJ. Prediction of composition dependency of glass forming ability of Mg-Cu-Y alloys by thermodynamic approach. *Scripta Materialia* (2005) 52(10):969–972.

Kim JH, Park JS, Lim HK, Kim WT, Kim DH. Heating and cooling rate dependence of the parameters representing the glass forming ability in bulk metallic glasses. *Journal of Non-Crystalline Solids* (2005) 351(16-17):1433-1440.

Kühn U, Eckert J, Mattern N, Schultz L. ZrNbCuNiAl bulk metallic glass matrix composites containing dendritic bcc phase precipitates. *Applied Physics Letters* (2002) 80(14):2478-2480.

Laws KJ, Gun B, Ferry M. Influence of casting parameters on the critical casting size of bulk metallic glass. *Metallurgical and Materials Transactions A* (2009) 40(10):2377-2387.

Lu ZP, Liu CT. A new glass-forming ability criterion for bulk metallic glasses. *Acta Materialia* (2002) 50(13):3501-3512.

Li X, Bian X, Hu L, Wu Y, Guo J, Zhang J. Glass transition temperature of bulk metallic glasses: A linear connection with the mixing enthalpy. *Journal of Applied Physics* (2007) 101(10):103540.

Li X, Song K, Wu Y, Ji H, Wang L. The mismatch entropy for bulk metallic glasses: A thermodynamic approach. *Materials Letters* (2013) 107:17-19.

Lu ZP, Tan H, Ng SC, Li Y. The correlation between reduced glass transition temperature and glass forming ability of bulk metallic glasses. *Scripta Materialia* (2000) 42(7):667-673.

Lee SW, Lee SC, Kim YC, Fleury E, Lee JC. Design of a bulk amorphous alloy containing Cu–Zr with simultaneous improvement in glass-forming ability and plasticity. *Journal of Materials Research* (2007) 22(2):486-492.

Li J, ChenTC, Zekiy AO. Correlative study between elastic modulus and glass formation in ZrCuAl (X) amorphous system using a machine learning approach. *Applied Physics A* (2021) 127:720.

Liu X, Li X, He Q, Liang D, Zhou Z, Ma J, Yang Y, Shen J. Machine learning-based glass formation prediction in multicomponent alloys. *Acta Materialia* (2020) 201:182–190.

Lundberg SM, Lee SI. A unified approach to interpreting model predictions. *Advances in neural information processing systems* (2017) 30: 4768–4777.

Long Z, Wei H, Ding Y, Zhang P, Xie G, Inoue A. A new criterion for predicting the glass-forming ability of bulk metallic glasses. *Journal of Alloys and Compounds* (2009) 475(1–2):207–219.

Long T, Long Z, Pang B, Li Z, Liu X. Overcoming the challenge of the data imbalance for prediction of the glass forming ability in bulk metallic glasses. *Materials Today Communications* (2023) 35:105610.

Long T, Long Z, Peng Z. Rational design and glass-forming ability prediction of bulk metallic glasses via interpretable machine learning. *Journal of Materials Science*, (2023) *58*(21):8833–8844.

Liu G, Sohn S, Kube SA, Raj A, Mertz A, Nawano A, Gilbert A, Shattuck MD, O'Hern CS, Schroers J. Machine learning versus human learning in predicting glass-forming ability of metallic glasses. *Acta Materialia,* (2023) 243:118497.

Miracle DB, Senkov ON, Topological criterion for Metallic Glass Formation. *Materials Science and Engineering*: A (2003) 347(1-2):50-58.

Mukherjee S, Schroers J, Zhou Z, Johnson WL, Rhim W-K. Viscosity and specific volume of bulk metallic glass-forming alloys and their correlation with glass forming ability. *Acta Materialia* (2004) 52(12):3689-3695.

Makino A, Kubota T, Chang C, Makabe M, Inoue A. FeSiBP bulk metallic glasses with unusual combination of high magnetization and high glass-forming ability. *Materials Transaction* (2007) 48(11):3024-3027.

Malekan M, Rashidi R, Bozorg M, Birbilis N. Tailoring the glass forming ability, mechanical properties and corrosion resistance of Cu–Zr–Al bulk metallic glasses by yttrium addition. *Intermetallics* (2023) 158:107906.

Ma H, Xu J, Ma E. MG-based bulk metallic glass composites with plasticity and high strength. *Applied Physics Letters* (2003) 83(14):2793-2795.

Mastropietro DG, Moya JA. Design of Fe-based bulk metallic glasses for maximum amorphous diameter (Dmax) using machine learning models. *Computational Materials Science* (2021) 188:110230.

Majid A, Ahsan SB, Tariq NUH. Modeling glass-forming ability of bulk metallic glasses using computational intelligent techniques. *Applied Soft Computing Journal* (2015) 28:569–578.

Nishiyama N, Takenaka K, Miura H, Saidoh N, Zeng Y, Inoue A. The world's biggest glassy alloy ever made *Intermetallics* (2012) 30:19-24.

Prabhu Y, Vincent S, Bhatt J. Thermodynamic modeling to optimize glass forming composition in the multicomponent Zr-cu-Co-Al System. *Materials Today: Proceedings* (2020) 28:1239-1244.

Ribeiro MT, Singh S, Guestrin C. Why should I trust you? Explaining the predictions of any classifier. *Proceedings of the 22nd ACM SIGKDD international conference on knowledge discovery and data mining*. 2016.

Reddy G, Saboo T, Ayyagari KPR. Prediction of glass forming ability of bulk metallic glasses using machine learning. *Integrating Materials and Manufacturing Innovation* (2021) 10:610-626.

Steinberg J, lord AE, Lacy LL, Johnson J. Production of bulk amorphous $Pd_{77.5}Si_{16.5}Cu_6$ in a containerless low-gravity environment. *Applied Physics Letters* (1981) 38(3):135-137.

Shao L, Dayte A, Huang J, Ketkaew J, Sohn SW, Zhao S, Wu S, Zhang Y, Schwarz UD, Schroers J. Pulsed laser beam welding of $Pd_{43}Cu_{27}Ni_{10}P_{20}$ bulk metallic glass. *Scientific Reports* (2017) 7(1):7989.

Suryanarayana C, Seki I, Inoue A. A critical analysis of the glass-forming ability of alloys. *Journal of Non-Crystalline Solids* (2009) 355(6):355-360.

Samavatian M, Gholamipour R, Samavatian V. Discovery of novel quaternary bulk metallic glasses using a developed correlation-based neural network approach. *Computational Materials Science* (2021) 186:110025

Sun, YT, Bai HY, Li MZ, Wang WH. Machine learning approach for prediction and understanding of glass-forming ability. *Journal of Physical Chemistry Letters* (2017) 8(14):3434-3439.

Suryanarayana, C, Inoue A. *Bulk Metallic Glasses*. CRC Press, 2017.

Turnbull D. Under what conditions can a glass be formed? *Contemporary Physics* (1969) 10(5):473-488.

Tan H, Zhang Y, Li Y. Synthesis of la-based in-situ bulk metallic glass matrix composite. *Intermetallics* (2002) 10(11-12):1203-1205.

Tripathi MK, Ganguly S, Dey P, Chattopadhyay PP. Evolution of glass forming ability indicator by genetic programming. *Computational Materials Science* (2016) 118: 56–65.

Vincent S, Murty BS, Bhatt J. Thermodynamic criteria for bulk metallic glass formation in Zr rich quaternary system. *AIP Conference Proceedings* (2012) 1447:583-584.

Wu W, Jiang J, Li G, Fu J, Jiang H, Gou P, Zhang L, Liu W and Zhao J. Ultrasonic additive manufacturing of bulk Ni-based metallic glass. *Journal of Non-Crystalline Solids* (2019) 506:1-5.

Wu W, Liu W, Du H, Wang B, Li G, Sun B, Zhang S and Zhao J. Optimization of sintering time and holding time for 3D printing of Fe-based metallic glasses. *Metals* (2018) 8(6):429.

Williams E, Lavery N. Laser processing of Bulk Metallic Glass: A Review. *Journal of Materials Processing Technology* (2017) 247:73-91.

Ward L, Agrawal A, Choudhary A, Wolverton C. A general-purpose machine learning framework for predicting properties of inorganic materials. *Npj Computational Materials* (2016) 2:16028.

Ward L, O'Keeffe SC, Stevick J, Jelbert GR, Aykol M, Wolverton C. A machine learning approach for engineering bulk metallic glass alloys. *Acta Materialia* (2018) *159*:102–111.

Wang DJ, Huang YJ, Wu LZ, Shen J. Mechanical behaviors of diamond reinforced TI-based bulk metallic glassy composites prepared by Spark Plasma Sintering. *Materials Science and Engineering: A* (2013) 560:841-846.

Xu Y, Huang X, Wang W. Preparation of bulk metallic glass $Pd_{40}Ni_{40}P_{20}$ under high pressure. *Applied Physics Letters* (1990) 56(20):1957-1958.

Xiong J, Shi SQ, Zhang TY. Machine learning prediction of glass-forming ability in bulk metallic glasses. *Computational Materials Science* (2021) 192:110362.

Zhou Z, Shang Y, Liu X, Yang Y. A generative deep learning framework for inverse design of compositionally complex bulk metallic glasses. *Npj Computational Materials* (2023) *9*(15):1-8.

Chapter 6

Deformation Behavior and Mechanisms in Metallic Glasses: Insights from an Atomistic Scale

Natraj Yedla*, **PhD**

Computational Materials Engineering Group,
Department of Metallurgical and Materials Engineering,
National Institute of Technology Rourkela, India

Abstract

Metallic glasses are one of the advanced structural materials having disordered atomic structures. These materials possess high strength and low ductility in comparison to their crystalline counterparts. The commonly observed deformation mechanism in metals, such as dislocation slip, twinning, or diffusion, is not observed in metallic glasses. The atomic level structure can be a combination of disordered atomic clusters plus the empty space between them, which is terms as free volume. It is the free volume region in which the atomic rearrangements occur due to the application of temperature or stress and results in plastic deformation. To date, there are two proposed deformation mechanisms in metallic glasses, i.e., the diffusive movement of atoms, and the second is the shearing of shear transformation zones (STZs). Furthermore, shear bands, which are an aggregate of STZs, nucleate and propagate to cause plastic deformation. In the above case, the deformation is observed to be heterogeneous. The following sections will present a detailed introduction, classical molecular dynamics methodology, and the deformation behavior and mechanisms in metallic glasses at the nanoscale.

* Corresponding Author's Email: yedlan@nitrkl.ac.in.

In: Properties and Uses of Metallic Glass
Editor: Shiv Prakash Singh
ISBN: 979-8-89113-569-7

Keywords: metallic glasses, molecular dynamics, free volume, shear bands

Introduction

The first reported metallic glass was Au75Si25 alloy in the form of a foil which was produced at Caltech, Pasadena, California by W. Klement (Jr.), Willens, and Duwez in the year 1960 (Klement et al., 1960). The cooling rate was of the order of 10^5 K/s–10^6 K/s in order to avoid crystallization. Research and developments lead to the production of multi-component metallic glasses in the bulk form (thickness ~10 mm) in Fe-based, Zr-based, Ti-based, Cu-based, Pd-based, etc. (Inoue, 1997; Inoue et al., 2008) systems. The tensile and compression tests studies on Cu and Zr-based metallic glasses show that the yield strength varied from 1 GPa–2 GPa (Inoue, 2008). The plastic strain in these materials is found to increase with the Al content and also the presence of a crystalline phase (Das et al., 2005). The fractographic analysis revealed vein like patterns on the fracture surface and shear bands on the gauge surface of the specimens (Lee et al., 2006; Wei et al., 2009; Sergueeva et al., 2005). Further, these bands are observed to be oriented at an angle to the loading direction. The mechanisms for the formation of the above patterns is still unclear but are assumed to be formed due to melting (Sergueeva et al., 2004; Sergueeva et al., 2005) in the former case and due to atomic rearrangements in the latter (Schuh and Lund, 2003; Schuh et al., 2007). Several quantitative models and atomistic simulations using molecular dynamics (MD) have been used to understand the deformation mechanisms in metallic glasses (Argon, 1979; Schuh and Lund, 2003). Figure 1 shows the atomic snapshot of the crystalline (Figure 1a) and glassy structure (Figure 1b) of a metallic system. The glassy structure is obtained by rapid cooling at a cooling rate of 10^{11} K/s.

Dislocation slip, twinning, and diffusion are the observed deformation mechanisms in crystalline materials. However, due to the disordered atomic structure, the deformation mechanism in metallic glasses is different. Despite lacking long-range order, metallic glasses possess short-range order, which is observed in several atomic packing models (Saksl et al., 2003; Schenk et al., 2004; Hirata et al., 2010; Lee et al., 2011). In glasses, the total volume can be divided into the space occupied by the atomic clusters and the empty space between them. The empty space is termed free volume (Turnbull and Cohen, 1960; Turnbull and Cohen, 1970).

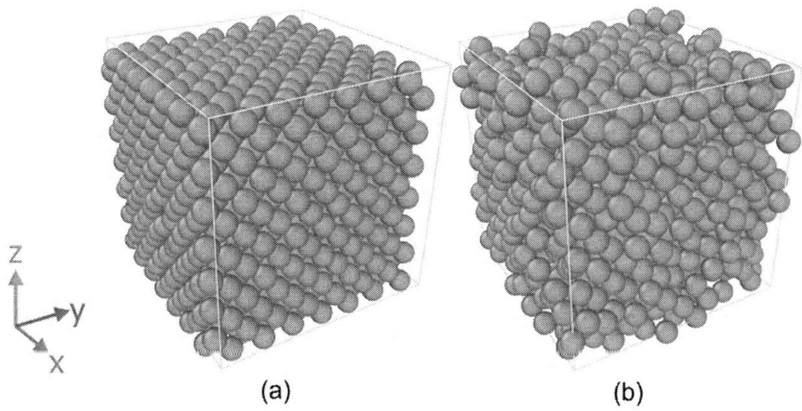

Figure 1. Atomic snapshots of the crystalline (a) and glassy structures (b).

The exact nature of local atomic motion during deformation is not fully resolved (Schuh et al., 2007) due to the lack of long-range ordering. However, there is an understanding that local atomic rearrangements lead to shear strain (Schuh et al., 2007). There are two theoretical models that are proposed to explain the deformation mechanism in metallic glasses, i.e., the free volume model (Spaepen, 1977) and shear transformation (Argon, 1979). Spaepen (1977), based on the inspection of the samples and fracture surfaces, had proposed two basic modes of deformation in metallic glasses: homogeneous flow and inhomogeneous flow.

1. Homogeneous flow

- Occurs at low stresses and high temperature
- Each volume element contributes to strain.
- Fracture occurs after extensive plastic flow, when certain regions of the specimen had thinned down to zero thickness.

2. Inhomogeneous flow

- Inhomogeneous flow occurs at high stress levels and low temperature.
- The stress is insensitive to strain rate.
- The flow is localized in to shear bands which indicate local softening of the material. There is some structural change in these bands.

- Fracture occurs along the plane which makes an angle of 45° with the tensile axis.
- Vein patterns are observed on the fracture surface.

The state of a glass is determined by its viscosity (η), and is one of the macroscopic key observables. Viscosity determines the flow behavior of glass which varies with temperature regime, i.e., above the melting point ($>T_m$), super cooled liquid region ($>T_g$ and $<T_m$) and below the glass transition temperature ($<T_g$) and hence, governs the mechanical properties of glass. The microscopic mechanism which governs both homogeneous and inhomogeneous flow is the atomic jump and macroscopic flow occurs as a result of a large number of individual atomic jumps (Spaepen (1977)). For an individual atomic jump there should be nearest neighbor environment, i.e., there must be a hole large enough to accommodate its large volume, and hence a potential jump site. Spaepen (1977) proposed that, at higher stresses free volume can be created and the lowering of the viscosity in shear bands must be due to an increase in free volume. Further, a small number of diffusive jumps lead to annihilation of free volume.

The first quantitative model of shear transformation zone (STZ) was developed by Argon (1979). At high temperatures $>T_g$ and low stress large plastic strain was mainly due to rearrangement of atoms in the regions around free volume sites and at high stress and low temperature the shear transformations become more intense and localized to few rows of atoms around a free volume site. The following assumptions were made

1. The bonds are largely metallic
2. The shear resistance across a plane resembles that of a close packed plane as in metals.

Shear transformation zone (STZ), is a small cluster of randomly close-packed atoms that spontaneously and cooperatively reorganize under the action of an applied shear stress as shown in Figure 2 and results in strain (Schuh and Lund, 2003) and was considered as the fundamental unit of plastic deformation in metallic glasses (Argon, 1979; Schuh and Lund, 2003; Schuh et al., 2007; Spaepen, 1977). As stated by Schuh et al., (2007) an STZ cannot be identified in a structure, however, it can be detected from the change in the structure due to straining. Further, the above theories suggested that STZ operation was strongly influenced by the local atomic arrangements and hence influences the structural evolution during deformation. The expression for free energy for STZ activation in terms of elastic constants of the glass is given in Ref. Schuh et al. 2007. The following are the features of an STZ which are summarized form the above theories:

1. STZs initiate around free-volume sites under applied shear stress due to high elastic strain at these sites.
2. The size of an STZ is about 1 nm - 2 nm comprising few atoms to 100 atoms.
3. STZs structure, size and energy scales vary from one glass to another.

The yield criterion in polycrystalline materials is given by von Mises criterion and Tresca criterion, but metallic glasses follow Mohr-Coulomb criterion (Donovan, 1989). It includes both the normal stress component on the slip plane and the applied shear stress. The validity of Mohr-Coulomb criterion was determined by conducting several mechanical tests. The planes of maximum resolved shear stress were found to deviate from 45° to 41.9° in the deformed metallic glasses and hence obey Mohr-Coulomb criterion

Classical Molecular Dynamics Simulation

Classical Molecular dynamics (MD) is a computer simulation method for investigating the motion of atoms and molecules (Yip, 2007). The atoms and molecules are assumed to obey classical Newtonian dynamics, and the interaction between the atoms/molecules is described by a force field. The typical timestep in MD is in the order of 1 femtosecond to avoid discretization errors. The flow chart for the MD code execution is shown in Figure 2.

LAMMPS is a classical molecular dynamics open-source code (Thompson et al., 2022) developed by Sandia Labs. The interatomic potentials available in LAMMPS are Lennard-Jones, Buckingham, Morse, EAM, Finnis/Sinclair EAM, modified EAM (MEAM), embedded ion method (EIM), Stillinger-Weber, Tersoff, REBO, AIREBO, and ReaxFF. The different ensembles available in LAMMPS are NVE (number of atoms: N, volume: V, and energy: E are constant), NPT (number of atoms: N, pressure: P, and temperature: T are constant), and NVT (number of atoms: N, volume: V, and temperature: T are constant). The boundary conditions that are used in MD are periodic (P) and non-periodic (S).

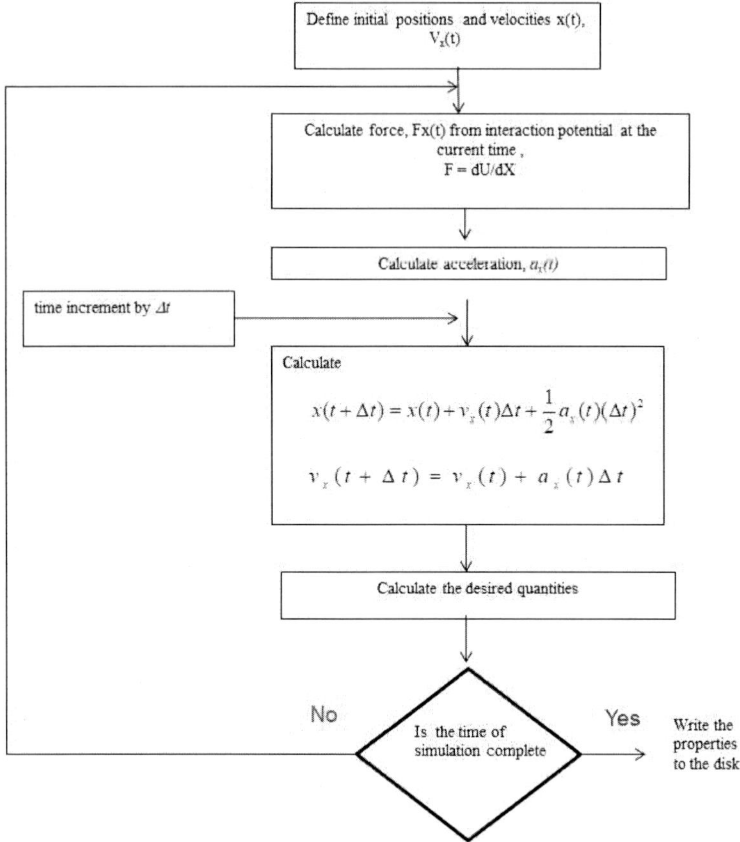

Figure 2. Flowchart for MD code execution.

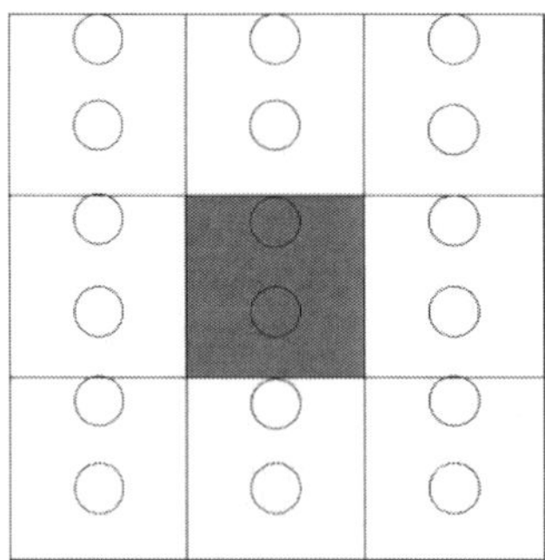

Figure 3. Schematic of a periodic boundary.

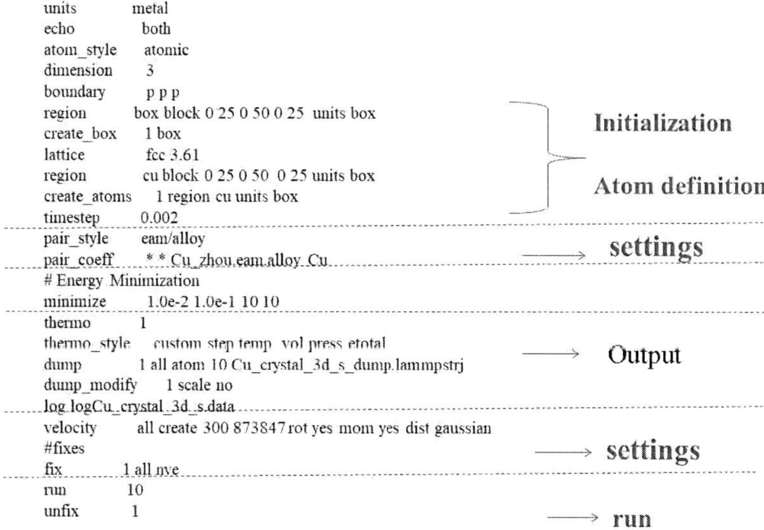
```
units          metal
echo           both
atom_style     atomic
dimension      3
boundary       p p p
region         box block 0 25 0 50 0 25  units box
create_box     1 box
lattice        fcc 3.61
region         cu block 0 25 0 50  0 25 units box
create_atoms   1 region cu units box
timestep       0.002
pair_style     eam/alloy
pair_coeff     * * Cu_zhou.eam.alloy Cu
# Energy Minimization
minimize       1.0e-2 1.0e-1 10 10
thermo         1
thermo_style   custom step temp  vol press etotal
dump           1 all atom 10 Cu_crystal_3d_s_dump.lammpstrj
dump_modify    1 scale no
log logCu_crystal_3d_s.data
velocity       all create 300 873847 rot yes mom yes dist gaussian
#fixes
fix            1 all nve
run            10
unfix          1
```


Figure 4. Example of a MD input script showing the commands and the syntax. Also, the several sections are indicated.

Figure 3 shows the schematic of a periodic boundary. The grey color region is the primary cell which has no free surfaces. The periodic boundary condition is used to eliminate free surfaces and simulate a large system by modeling a small part (primary cell; yellow color) that is far from its edge. The integrators that are available in LAMMPS are velocity-Verlet integrator, Brownian dynamics, and energy minimization via conjugate or steepest descent relaxation. Example of a MD input script showing the commands and the syntax is shown in Figure 4.

Microstructure and Its Evolution During Deformation

Diffraction methods have limitations in detecting the ordering in metallic glasses (Gokularatnam, 1974). The structural models suggested that amorphous alloys contain various types of ordered atomic clusters viz. SRO (short range order) which was first evidenced by Miracle (2004) in his cluster-packing model. Furthermore, Sheng et al., (2006) proposed that short to medium range ordering (MRO) exists in model Ni80P20 metallic glass. Whilst ISRO (icosahedral short range order) was found in Zr70Cu30, Zr70Cu29Pd1 supercooooled liquids (Saksl et al., 2003) by x-ray absorption fine structure. In

Al-Fe-Co supercooled melts CSRO (chemical short range order) of preferentially Al atoms was observed (Schenk et al., 2004) where the ordering enhanced with decrease in temperature. These structures affect the atomic packing state and hence may influence the properties of glasses.

In the ab-initio MD simulation studies on the atomic configurations of liquid and glassy Mg65Cu25Y10 alloys (Hui et al., 2008), it was shown that short-to-medium range icosahedral ordering exists. It was suggested that the presence of these structures made the under cooled melt stable and improved the GFA. The presence of short-range order (SRO) and medium-range order (MRO) clusters is observed in metallic glasses (Yedla and Ghosh, 2017). It has been reported that the Zr-centered clusters lower the diffusivity in Cu-Zr metallic glasses (Peng et al., 2010). In CuZr Al metallic glasses, Cu and Al-centered icosahedral clusters are identified as the local structural motifs. Interpenetrating connection of icosahedra (ICOI) clusters and network in $Cu_{50}Zr_{50}$ and $Cu_{65}Zr_{35}$ MGs have been identified using common neighbor analysis (CNA) (Lee et al., 2011). The high strength of $Cu_{65}Zr_{35}$ metallic glass has been attributed to the large fractions of ICOI clusters connected over a network. Further, it is reported that the atomic mobility increases with the disruption of ordered clusters in CuZr metallic glasses (Zhang et al., 2014).

Several loading types, such as tensile, compression, shear, indentation, and bending, are carried out to understand the deformation behavior and mechanism in metallic glasses. Schuh and Lund, 2003 propose that the amorphous metals plastically yield following the Mohr–Coulomb criterion. Cheng and Ma, 2011 studies by shear tests on defect free metallic glasses report that shear bands nucleate homogeneously, which is in contradiction to that observed in experiments. Li and Li, 2007 have shown that shear bands initiate at the notch tips and grow. Further, the nucleation zone size is found to be 10-20 nm. Wang et al., 2010 studies showed that shear transformation zones and shear bands are observed during nanoindentation and compression studies of defect metallic glasses. The sites of pores act as stress concentrators and greatly promote the initialization and propagation of shear bands. Feng et al., 2015 report a planar like type of <0, 2, 8, 5> network in shear bands in $Cu_{64}Zr_{36}$ metallic glasses subjected to tensile tests. Rejuvenation or structural relaxation has been reported in Zr-based metallic glasses due to the weakening of medium-range order clusters (Feng et al., 2018). Tao et al., 2021 propose that rejuvenation will induce more nucleation sites for STZs, and physical aging will decrease the liquid like zones in the deformation studies on Zr-based metallic glasses. Katakareddi and Yedla, 2022 in the deformation studies on Cu-Zr metallic glasses, report the presence of atomic clusters of large shear strain at the yield as compared to the bulk, and these clusters develop into bands with increasing plastic deformation. Figure 5 shows the tensile stress-strain curve of Cu-Zr metallic glass combined with the atomic snapshots of shear strain analysis (Katakareddi and Yedla, 2023). The regions with high shear strain increase with the deformation, and after the peak stress, strain localization is observed (e = 0.2)

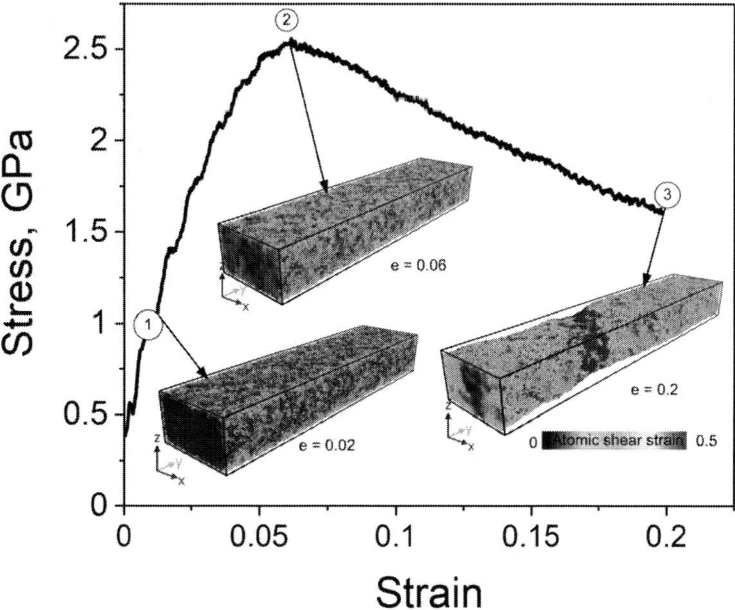

Figure 5. Stress-strain curve of Cu-Zr metallic glass and the inset shows the atomic snapshots with shear strain distribution in the specimen corresponding to different regions on the stress-strain curve.

Conclusion

Metallic glasses are metallic materials with disordered atomic structures. The atomic level structure is a combination of disordered atomic clusters and the free volume which is the empty space between them. The deformation mechanism is different from the crystalline counterpart because of the disorderd structure. The plastic deformation is reported to occur in the regions of free volume where the atomic rearrangements occur due to the application of temperature or stress. Apart from diffusive movement of atoms, shearing of shear transformation zones (STZs) are the two deformation mechanism proposed. Also, it is reported that the propagation of shear bands cause plastic deformation. Several MD models have been used to validate the experimental observations.

References

Argon AS, Plastic-deformation in metallic glasses, *Acta Metallurgica*, (1979) 27(1):47-58.

Cheng YQ, Ma E. Intrinsic shear strength of metallic glass. *Acta Materialia* (2011) 59(4):1800-1807.

Das J, Tang MB, Kim KB, Theissmann R, Baier F, Wang WH, Eckert J. "Work-hardenable" ductile bulk metallic glass, *Physical Review Letters*, (2005) 94(20):205501.

Donovan PE. A yield criterion for $Pd_{40}Ni_{40}P_{20}$ metallic-glass, *Acta Metallurgica* (1989) 37(2): 445-456.

Feng SD, Chan KC, Zhao L, Pan SP, Qi L, Wang LM, Liu RP. Rejuvenation by weakening the medium range order in $Zr_{46}Cu_{46}Al_8$ metallic glass with pressure preloading: A molecular dynamics simulation study. *Materials & Design* (2018) 158:248-255.

Feng S, Qi L, Wang L, Pan S, Ma M, Zhang X, Li G, Liu R. Atomic structure of shear bands in $Cu_{64}Zr_{36}$ metallic glasses studied by molecular dynamics simulations. *Acta Materialia* (2015) 95:236-243.

Gokularathnam CV. Structure of metallic glasses, *Journal of Materials Science* (1974) 9(4): 673-682.

Peng HL, Li MZ, Wang WH, Wang CZ, Ho KM. Effect of local structures and atomic packing on glass forming ability in Cu_xZr_{100-x} metallic glasses, *Appl Phys Lett* (2010) 96:21901.

Hirata A, Guan PF, Fujita T, Hirotsu Y, Inoue A, Yavari AR, Sakurai T, Chen MW. Direct observation of local atomic order in a metallic glass, *Nature Materials* (2011) 10 (1): 28-33.

Inoue A. Bulk amorphous alloys with soft and hard magnetic properties. *Materials Science and Engineering: A* (1997) 15:226: 357-63.

Inoue A, Wang XM, Zhang W. Development and application of bulk metallic glasses, *Review of Advanced Materials Science* (2008) 18:1-9.

Katakareddi G, Yedla N. The effect of loading methods on the microstructural evolution and degree of strain localization in $Cu_{50}Zr_{50}$ metallic glass composite nanowires: A molecular dynamics simulation study. *Journal of Molecular Graphics and Modelling* (2022) 115:108216.

Katakareddi G, Yedla N. Investigation of strain localization in metallic glasses: size effects (Unpublished work) 2023.

Klement W, Willens RH, Duwez POL. Non-crystalline structure in solidified gold-silicon alloys, *Nature* (1960) 187(4740):869-870.

Lee M, Lee CM, Lee KR, Ma E, Lee JC. Networked interpenetrating connections of icosahedra effects on shear transformations in metallic glass, *Acta Materialia* (2011) 59(1):159-170.

Lee SW, Huh MY, Fleury E, Lee JC. Crystallization-induced plasticity of Cu-Zr containing bulk amorphous alloys, *Acta Materialia* (2006) 54 (2):349-355.

Li QK, Li M. Assessing the critical sizes for shear band formation in metallic glasses from molecular dynamics simulation. *Applied Physics Letters* (2007) 91(23):231905.

Miracle DB. A structural model for metallic glasses, *Nature Materials*, (2004) 3(10):697-702.

Saksl K, Franz H, Jovari P, Klementiev K, Welter E, Ehnes A, Saida J, Inoue A, Jiang JZ. Evidence of icosahedral short-range order in $Zr_{70}Cu_{30}$ and $Zr_{70}Cu_{29}Pd_1$ metallic glasses, *Applied Physics Letters* (2003) 83(19).3924-3926.

Schenk T, Simonet V, Holland-Moritz D, Bellissent R, Hansen T, Convert P, Herlach DM. Temperature dependence of the chemical short-range order in undercooled and stable Al-Fe-Co liquids, *Europhysics Letters* (2004) 65(1):34-40.

Sheng HW, Luo WK, Alamgir FM, Bai JM, Ma E. Atomic packing and short-to-medium-range order in metallic glasses, *Nature* (2006) 439 (7075):419-425.

Schuh CA, Hufnagel TC, Ramamurty U. Mechanical behavior of amorphous alloys, *Acta Materialia* (2007)55 (12):4067-4109.

Schuh CA, Lund AC, Atomistic basis for the plastic yield criterion of metallic glass, *Nature Materials* (2003) 2(7):449-452.

Sergueeva AV, Mara NA, Branagan DJ, Mukherjee AK. Strain rate effect on metallic glass ductility, *Scripta Materialia* (2004) 50(10):1303-1307.

Sergueeva AV, Mara NA, Kuntz JD, Lavernia EJ, Mukherjee AK. Shear band formation and ductility in bulk metallic glass, *Philosophical Magazine* (2005) 85 (23):2671-2687.

Spaepen F. A microscopic mechanism for steady state inhomogeneous flow in metallic glasses, *Acta Metallurgica* (1977) 25(4):407-415.

Tao K, Qiao JC, He QF, Song KK, Yang Y. Revealing the structural heterogeneity of metallic glass: mechanical spectroscopy and nanoindentation experiments. *International Journal of Mechanical Sciences* (2021)201: 106469.

Thompson AP, Aktulga HM, Berger R, Bolintineanu DS, Brown WM, Crozier PS, Plimpton SJ. LAMMPS-a flexible simulation tool for particle-based materials modeling at the atomic, meso, and continuum scales. *Computer Physics Communications* (2022) 271:108171.

Turnbull D, Cohen MH. On the free-volume model of the liquid-glass transition, *The Journal of Chemical Physics* (1970) 52(6):3038-3041.

Turnbull D, Cohen MH. Crystallization kinetics and glass formation, in: Modern Aspects of the Vitreous State, Ed. Mackenzie, J. D., London: Butterworth, First Edition, (1960): 38-62.

Wang J, Hodgson PD, Zhang J, Yan W, Yang C. Effects of pores on shear bands in metallic glasses: A molecular dynamics study. *Computational materials science* (2010) 50(1):211-217.

Wei XF, Sun YF, Guan SK, Terada D, Shek CH. Compressive and tensile properties of CuZrAl alloy plates containing martensitic phases, *Materials Science and Engineering A* (2009) 517(1-2):375-380.

Yip S. (Ed.). *Handbook of materials modeling*. Springer Science & Business Media, 2007.

Yedla N, Ghosh S. Nature of atomic trajectories and convective flow during plastic deformation of amorphous $Cu_{50}Zr_{50}$ alloy at room temperature-classical molecular dynamics studies. *Intermetallics* (2017) 80:40-47.

Zhang Y, Mendelev MI, Wang CZ, Ott R, Zhang F, Besser MF, Ho KM, Kramer MJ. Impact of deformation on the atomic structures and dynamics of a Cu-Zr metallic glass: A molecular dynamics study. *Physical Review B* (2014) 90(17):174101.

Index

E

F

G

H

I

J

L

M

N

P

Q

R

S

T

U

V